SpringerBriefs in Petroleum Geoscience & Engineering

The SpringerBriefs series in Petroleum Geoscience & Engineering promotes and expedites the dissemination of substantive new research results, state-of-the-art subject reviews and tutorial overviews in the field of petroleum exploration, petroleum engineering and production technology. The subject focus is on upstream exploration and production, subsurface geoscience and engineering. These concise summaries (50–125 pages) will include cutting-edge research, analytical methods, advanced modelling techniques and practical applications. Coverage will extend to all theoretical and applied aspects of the field, including traditional drilling, shale-gas fracking, deepwater sedimentology, seismic exploration, pore-flow modelling and petroleum economics. Topics include but are not limited to:

- Petroleum Geology & Geophysics
- Exploration: Conventional and Unconventional
- Seismic Interpretation
- Formation Evaluation (well logging)
- Drilling and Completion
- Hydraulic Fracturing
- Geomechanics
- Reservoir Simulation and Modelling
- Flow in Porous Media: from nano- to field-scale
- Reservoir Engineering
- Production Engineering
- Well Engineering; Design, Decommissioning and Abandonment
- Petroleum Systems; Instrumentation and Control
- Flow Assurance, Mineral Scale & Hydrates
- Reservoir and Well Intervention
- Reservoir Stimulation
- Oilfield Chemistry
- Risk and Uncertainty
- Petroleum Economics and Energy Policy

Contributions to the series can be made by submitting a proposal to the responsible Springer contact, Anthony Doyle at anthony.doyle@springer.com.

Reza Yousefzadeh · Alireza Kazemi ·
Mohammad Ahmadi · Jebraeel Gholinezhad

Introduction to Geological Uncertainty Management in Reservoir Characterization and Optimization

Robust Optimization and History Matching

 Springer

Reza Yousefzadeh
Department of Petroleum Engineering
Amirkabir University of Technology
Tehran, Iran

Mohammad Ahmadi
Department of Petroleum Engineering
Amirkabir University of Technology
Tehran, Iran

Alireza Kazemi
Department of Petroleum and Chemical
Engineering, College of Engineering
Sultan Qaboos University
Muscat, Oman

Jebraeel Gholinezhad
School of Energy and Electronic
Engineering
University of Portsmouth
Portsmouth, UK

ISSN 2509-3126   ISSN 2509-3134 (electronic)
SpringerBriefs in Petroleum Geoscience & Engineering
ISBN 978-3-031-28078-8   ISBN 978-3-031-28079-5 (eBook)
https://doi.org/10.1007/978-3-031-28079-5

This Springer imprint is published by the registered company Springer Nature Switzerland AG
The registered company address is: Gewerbestrasse 11, 6330 Cham, Switzerland

This book is dedicated:

To my wonderful friend, Meysam Yari, and my family that always supported me.

Reza Yousefzadeh—January 2023

Preface

The main purpose of this book is to investigate the methods, challenges, and solutions to manage uncertainty in reservoir characterization and optimization. This book is a quick and strong introduction to reservoir characterization and optimization under geological uncertainty that can guide graduate/post-graduate students and researchers to start their investigations with a strong understanding of the basics of geological uncertainty and solutions to it. The content of this book is prepared based on the most prominent works by a large number of authors including the authors of this book who have had several published works regarding the geological uncertainty management both in reservoir characterization and in optimization. Since this book is a brief, methods and examples are briefly described and some outdated methods are ignored.

This book contains five chapters. In Chap. 1, types and sources of uncertainty, uncertainty in different reservoir management phases, and challenges in those phases are introduced, and general methods to manage those uncertainties are explained. Chapter 2 is about the "Geological Uncertainty Quantification". This type of uncertainty is related to macro- (e.g., fluid contacts) and micro-scale (e.g., permeability distribution) parameters that affect all decisions about field development. Use of geological prior information, seismic and petrophysical data, geological parametrization, and exploring the range of scenarios are explained in this chapter. Then, geological realizations and methods to generate them are explained; their workflow, advantages, and disadvantages are discussed. Chapter 3 is dedicated to "Reducing the Geological Uncertainty by History Matching". History matching suffers from different challenges including high dimensionality of the models, nonlinear, non-Gaussian, and three-dimensional distribution of the uncertain parameters, which are introduced in this chapter. In the end, different approaches to history matching, including the open-loop and closed-loop reservoir management as well as the common methods of history matching, are described; their advantages and shortcomings are discussed. In Chap. 4, dimensionality reduction methods to resolve the challenges related to high dimensionality of the geological realizations are presented. Dimensionality reduction methods proposed by the authors of this book and others, including the conventional and novel machine learning-based methods, are presented, and their pros and cons are discussed. Chapter 5 is about "Field Development Optimization Under Geological Uncertainty" using robust optimization. Challenges in robust field development optimization, including the high computational cost, risk attitudes, and choosing a subset of realizations to use in robust optimization, are discussed, and common solutions are introduced. In this regard, recent methods of selecting a subset of realizations using static, dynamic, and hybrid properties of the reservoir in green and brown fields and novel approaches to constraining the search space of the optimizers to facilitate the robust optimization process proposed by the authors and other researchers are presented. In the last chapter (Chap. 6), different kinds of proxy models in history matching and robust field development optimization are introduced. These proxies include the physics-based, non-physics-based, and hybrid proxies. Their pros and cons are discussed, and some of their applications in history matching and robust optimization are presented.

Tehran, Iran
Muscat, Oman
Tehran, Iran
Portsmouth, UK

Reza Yousefzadeh
Alireza Kazemi
Mohammad Ahmadi
Jebraeel Gholinezhad

Acknowledgements As the first author, I am greatly thankful to Dr. Mohammad Sharifi (Associate Professor at Amirkabir University of Technology) who always inspired me to broaden my knowledge in various fields of study. I am also thankful to Dr. Hamidreza M. Nick (Senior Researcher at Danish Offshore Technology Centre) who inspired me to do great works.

In the end, the authors are thankful to all friends, family, colleague, and collaborators for their support, feedback, inspiration, and discussions that helped us to prepare and improve this book.

Contents

List of Figures

Chapter 1
Uncertainty Management in Reservoir Engineering

Abstract Almost all activities in real life entail different kinds of uncertainty. From daily decisions to complicated problems, such as petroleum reservoir characterization, suffer from uncertainties. Uncertainty can have different roots, including incomplete observation of the system, incomplete modeling of the system because of our limited knowledge and understanding of the underlying mechanisms and rules of the system, intrinsic uncertainty in the system, and inaccurate measurement of the system's parameters. The first step to deal with uncertainty is to recognize its root and type. This chapter introduces different types of uncertainty in reservoir engineering, challenges induced by uncertainty, and common ways to treat them. Two general approaches to handle the uncertainty are described, named forward uncertainty management and inverse uncertainty management. Forward uncertainty management tries to propagate the uncertainty from inputs to the output(s) to make robust decisions regarding the problem understudy. On the other hand, inverse uncertainty management deals with calibrating model parameters to reduce the range of uncertainty.

Keywords Uncertainty · Reservoir engineering · History matching · Robust optimization · Bayesian framework · Geological uncertainty · Field development

1.1 Introduction to Uncertainty Management in Reservoir Engineering

Uncertainty is present in almost all kinds of activities and processes. In fact, except for simple mathematical expressions, it is very hard to find an event that is completely true or that will happen deterministically [1]. This uncertainty increases the complexity of problems. For example, if a decision is made under some assumptions without considering the uncertainties, it may fail if the conditions turn out to be different from that of thought to be. This can yield instability in the decision making process. As a result, any decision making process should consider the uncertain nature of events, and the decisions should be made under the present uncertainties. Consequently, the decision will be stable if the situation changes.

Reservoir engineering problems are not exceptional from uncertainties. A definition to reservoir management under uncertainty is to reduce the range of uncertainty, or to make robust decisions about the field development optimization (FDO), or to make flexible decisions regarding the FDO that can be changed easily if our knowledge of the reservoir or the available technology to develop the field is upgraded [2–4]. Reservoir engineering under uncertainty is the process of calculating probabilistic outcomes of any decision making process under different sources of uncertainty. In other words, uncertainty can be taken into account by evaluating the performance of different activities using the probabilistic measures over numerous possible situations or models. The probabilistic measures can be the expected value of the quantities of interest with a risk attitude. Some common risk measures are the value at risk (VaR), conditional value at risk (CVaR), mean-semi-variance, mean-standard deviation, etc., which will be covered in Chap. 5.

Model-based development in reservoir engineering is one of the common practices in which future development plans are envisaged based on specific models. However, these models can be of uncertain nature. For example, uncertainty can be in the reservoir model itself, in market, in surface facility, etc. These uncertainties can be because of lack of information about the above parameters. Uncertainty in the reservoir model is the most challenging type of uncertainty in reservoir engineering that yields many difficulties in characterizing the reservoir model and predicting its future behavior. For example, In the deterministic modeling, the reservoir model is considered certain; however, in the probabilistic modeling, the reservoir model is uncertain and multiple reservoir models are used to account for the uncertainty.

To have an uncertainty management accounting for all relevant uncertainties, first, the sources of uncertainties should be identified. Then, the range of uncertainty (probability distribution) of the uncertain parameters should be characterized. After characterizing the uncertainty range, these uncertainties should be quantified by sampling from their distribution space. Then, these uncertainties have to be propagated through the model by evaluating the various configurations of the uncertain parameters. Finally, the effect of the uncertainties on the quantities of interest should be evaluated, and a decision which is stable over the range of uncertainty should be made.

Reservoir management under uncertainty is done in two ways: (1) One way is to reduce the uncertainty by incorporating more information in the models to shrink the range of uncertainties. (2) Another way is to make robust decisions by considering multiple realizations of the uncertain parameters in which the quantities of interest are calculated over different models and a robust and stable decision is made, which is optimal over all available realizations in terms of a specific objective function.

In this chapter, uncertainty in reservoir engineering, types and sources of uncertainty, uncertainty in different phases of field development, challenges in reservoir management under uncertainty, and the general uncertainty management methods are introduced. The first step in addressing uncertainty is to identify the type and source of the uncertainty.

1.2 Types and Sources of Uncertainty

In general, there are three sources of uncertainty that classifies the uncertainty into four categories:

1. **Intrinsic uncertainty in the system**: concepts in quantum mechanics and dynamics of sub-atomic particles are described probabilistically.
2. **Incomplete observations**: this type of uncertainty is related to our incomplete observation of the system variables even if the system is certain.
3. **Incomplete modeling**: if we ignore some portion of information in modeling, this will lead to uncertainty in the predictions of the model. For example, discretizing a continuous space into small cells results in uncertainty since the properties are assumed constant inside each cell.
4. **Measurement Error**: this kind of uncertainty arises because of limited resolutions and variability of the measurement tools. We can use repeated measurements to reduce the extend of this uncertainty.

In petroleum related problems, since our observation from underground reservoirs is limited, we cannot build a model that conforms completely with the true model. Consequently, this leads to an incomplete observation of the model parameters that is classified under the second category. The common approach to build reservoir models in reservoir engineering is history matching. In history matching, using the limited observations of the model parameters and dynamic data from a limited number of wells, it is tried to build a model that yields dynamic data with a minimum deviation from the observed data. Since the number of unknowns is quite higher than the number of knowns, history matching is an underestimate inverse problem. Consequently, there is an infinite number of reservoir models than can match the observed dynamic data, which leads to uncertainties in characterizing the reservoir. Also, because of our incomplete knowledge of the governing physical laws of fluid flow in porous media, we cannot build a perfect model of the governing laws in underground reservoirs. Therefore, the third type of uncertainty is also present in reservoir modeling problems. Furthermore, due to the limitations in the resolution of the measurement tools, the fourth type of uncertainty is also present in reservoir engineering problems.

1.2.1 Reservoir Characterization Uncertainty

Uncertainty in reservoir characterization is divided into two main groups: (1) geological uncertainties, (2) reservoir engineering uncertainties. Geological uncertainties are related to the lack of knowledge about the static reservoir properties (e.g., facies, permeability, porosity), structural framework (e.g., boundaries of the reservoir, layering, thickness), and discontinuities (e.g., existence of faults or fracture networks) [5–8]. These properties are represented by spatial distributions and do not refer to properties influenced by rock and fluid interactions (e.g., wettability, relative permeability, capillary pressure). The second group is the reservoir engineering uncertainties which refer to the lack of knowledge about reservoir properties other than geological properties, such as relative permeability, fluid properties, capillary pressure, etc. [9, 10].

1.2.2   Economic Uncertainty

Economic uncertainty arises because of our limited knowledge of the future. This kind of uncertainty is pertained to volatilities in oil prices, operational expenditures, capital expenditures, embargo and sales issues, etc. [11]. Since the oil price is dependent on many variables, such as the dollar index, production costs, supply/demand, global events, energy crisis, etc., we cannot use a fixed oil price in our economic analysis of the development plan.

1.2.3   Operational Uncertainty

Operational uncertainty is mainly attributed to challenges in operations, such as failures and delays in production and injection plans. Mechanical or technical failures in facilities and equipment are examples of operational uncertainties (e.g., inflow control valves (ICV) or inflow control devices (ICD) of intelligent wells, packers, downhole/surface pumps, injection pumps, surface separators and production units). Issues related to well integrity (e.g., casing collapse and internal leaks), natural disasters, market availability issues (e.g., inadequate quantities of rigs, pumps, infrastructure, fluid volumes) are other examples of operational uncertainties [12].

1.2.4   Surface Facility Uncertainty

Surface facility uncertainty can also be categorized as the second type of uncertainty. Surface facility uncertainty is related to the capacity of the processing unit on the surface which is represented by the maximum production/injection capacity, rig availability, minimum economic constraints, etc. These factors can be changed because of changes in the development goals, environmental regulations, oil prices, etc., which are uncertain. The uncertainty in these parameters lead to uncertainty in surface facilities.

1.2.5 Information Reliability

Information reliability refers to uncertainty in measurements, which is the result of the low resolution of measurement devices. Therefore, any measurement of reservoir's dynamic (e.g., pressure and production data, well tests) and static properties (e.g., seismic data, permeability, porosity) are associated with measurement uncertainty.

1.2.6 Modeling Uncertainty

Uncertainty in modeling refers to simplifications or errors in the developed computational method, mathematics, algorithms, and equations to model the fluid flow in the reservoir [13–15].

Among the geological, economic, modeling, operational, measurement, and surface facility uncertainties, the first one is the most challenging type of uncertainty. Therefore, this book is dedicated to the investigation on this type of uncertainty. Geological uncertainty can be encountered in different phases of development, including the history matching and prediction phases. In the following subsection, Uncertainty Treatment in Different Phases of Field Development is discussed.

1.3 Uncertainty Management in Different Phases of Field Development

In reservoir engineering, uncertainty can be present in two phases: (1) History matching, (2) Prediction. Uncertainty in the history matching and prediction phases are treated differently which are described in the following.

1.3.1 Uncertainty Management in History Matching

History matching is an optimization problem in which the model parameters are updated iteratively to get a match between the simulated and observed dynamic (such as production or 4D seismic data) data from the reservoir. However, since our observation from the underground reservoir is limited, the history matching problem will be an underestimate inverse problem that has an infinite number of solutions. Thereby, it is common to generated numerous prior reservoir model realizations using the hard and soft data gathered from core samples, seismic data, geologic interpretations, etc., and then condition these realizations to production/4D seismic data to get more realistic models. As mentioned above, history matching is an optimization problem in which the objective function is the difference between the simulated and observed dynamic data. Equation 1.1 represents a sample objective function used in a history matching process.

$$J = \sum_{i=1}^{N_p} \sum_{j=1}^{N_t} w_i \left(\frac{x_i^{obs}\left(t^j\right) - x_i^{sim}\left(t^j\right)}{STD_i} \right)^2, \tag{1.1}$$

where N_p is the number of matching parameters (cumulative oil production, water-cut, etc.), N_t is the total number of samples (timesteps), w_i is the weight of the ith matching parameter, $x_i^{obs}\left(t^j\right)$ is the observed value of the ith parameter at the jth timestep, $x_i^{sim}\left(t^j\right)$ is the simulated value of the ith parameter at the jth timestep, and STD_i is the standard deviation of the ith matching parameter. The goal of history matching is to minimize Eq. (1.1). Therefore, uncertainty management in the history matching phase is, in fact, conditioning the prior reservoir models to dynamic data. In other words, the goal is to reduce the range of uncertainty in the model parameters [16–18].

1.3.2   Uncertainty Management in Field Development Optimization

In the development phase, the goal is to predict the future performance of the reservoir, such as average reservoir pressure, production/injection rates, water-cut, etc., and profitability of the project under different development strategies. The success of this assessment is to have a reliable reservoir and economic model. However, as mentioned above, the uncertainties associated with geological models makes it difficult to create a unique model for evaluating different development scenarios. Consequently, it is necessary to apply a method that can capture the range of uncertainty in the model parameters. A common practice to account for the geological uncertainty in the development phase is to calculate the quantities of interest (cumulative oil production, profitability, etc.) for different realizations of the uncertain parameters and make a robust decision about the development plan that remains stable if the model parameters change [3, 12, 19, 20]. Uncertainty in both the history matching and development phases brings about some challenges that are presented in the next section.

1.4 Challenges in Reservoir Management Under Uncertainty

Reservoir management under uncertainty involves several challenges both in the history matching and prediction phases. The most common challenges are the High Number of Parameters in History Matching, Dimension of the Reservoir Models in History Matching, Non-linear and non-Gaussian Distribution of the Reservoir Properties, High Computational Costs, and Initial Ensemble of Realizations in History Matching, which are described in the following.

1.4.1 High Number of Uncertain Parameters in History Matching

This challenge is faced in the history matching phase. As discussed in Sect. 1.1, history matching is an optimization problem where the model parameters are updated to get a match between the simulated and observed production data. If the number of uncertain parameters is large, the optimization problem gets more complicated and causes troubles in the optimization process. For example, a reservoir model may consist of hundreds of thousands to millions of grid blocks. If we have to update the permeability of all grid blocks, then we have to update a large number of quantities each time to get new production data. High number of uncertain parameters of the problem not only increases the complexity of the problem, but also raises difficulties in preserving the geologic realism of the reservoir model. This is because of the redundant information that may exist in the input of the history matching methods.

1.4.2 Dimension of the Reservoir Models in History Matching

Dimension of the reservoir models is the spatial dimension of the models (one-, two-, or three-dimensional models). This is of concern when updating the reservoir models during the history matching process. Since the history matching methods take the input as a one-dimensional vector, this causes the loss of information related to 2D and 3D dependencies between the parameters of the models. In other words, as parameters, such as permeability or porosity, have a 3D structure, converting them into a one-dimensional vector causes loss of higher dimensional information. Also, it is not possible to capture the boundary effects when converting the parameters to 1D vectors.

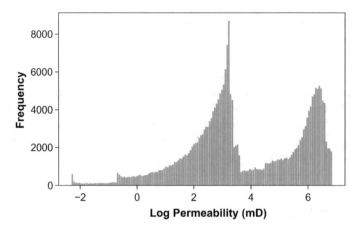

Fig. 1.1 Histogram of the logarithm of X permeability of a reservoir model

1.4.3   Non-linear and Non-Gaussian Distribution of the Reservoir Properties

Distribution of the reservoir model parameters, such as the permeability, usually follows a non-linear and non-Gaussian distribution. Figure 1.1 shows the histogram of the logarithm of X permeability of a reservoir model. As can be seen from this figure, the distribution is not Gaussian. However, some ensemble-based history matching methods (e.g., Ensemble Kalman Filter, Ensemble Smoother [21], see Sect. 3.6) work well on Gaussian inputs, but deteriorate on non-Gaussian inputs. Figure 4 in [21] shows the output of the Ensemble Smoother – Multiple Data Assimilation (ES-MDA) history matching method used to condition facies realizations to observed production data. As can be seen from Fig. 4a in [21], the output has a Gaussian distribution that requires some post-processing (Figs. 4b and c in [21]) to transform it into a Bernoulli distribution. This is also true for other ensemble-based methods such as Ensemble Kalman Filter (EnKF) as it has a similar formulation to ES-MDA with a Gaussian noise involved in its formulation. Therefore, a challenge in the history matching phase is to capture the non-Gaussian dependencies in the input realizations using some parameterization techniques.

1.4.4   High Computational Costs

One of the big challenges of reservoir management under uncertainty both in the history matching and the field development phases is the high computational cost associated with the process. History matching asks for numerous simulations of the reservoir model(s) to get the simulated data which are required for updating the reservoir model(s) during the data assimilation step. If the model has a large number of grid blocks, it will take a long time to get simulated data increasing the computational burden of the process. Therefore, the computational cost of the problem will be increased proportional to the number of reservoir model realizations.

1.4.5 Initial Ensemble of Realizations in History Matching

Ensemble-based history matching methods require an initial ensemble of realizations of the uncertain parameters to get conditioned to the production data. Random selection of these realizations may result in sampling from a limited region of the distribution that can deteriorate the overall performance of the history matching process especially if the realizations are sampled from a limited region with large deviations from the true model. Therefore, selecting an appropriate set of realizations as the initial ensemble is another challenge in history matching.

1.5 Methods of Reservoir Management Under Geological Uncertainty

Two main approaches can be taken into account to manage the geological uncertainty. These two approaches are the forward uncertainty management and inverse uncertainty management.

1.5.1 Forward Uncertainty Management

In this approach, the uncertainties are represented in the outputs of the system. Uncertainty is propagated from the uncertain input parameters to the output(s). This approach focuses on the analysis of uncertainty propagation targets:

- Calculating low-order statistics of the output. In other words, mean and variance.
- Analyzing the reliability of the output.
- Providing a probability distribution and statistical percentiles of the outputs.

Forward uncertainty management entails an uncertainty quantification step to sample a realization of the uncertain parameters to be used for obtaining the corresponding output of the system for that realization. Forward uncertainty quantification approaches are divided into probabilistic and non-probabilistic methods [22]. Some of the probabilistic forward uncertainty management approaches are listed below:

- Methods based on simulation: importance sampling, Monte-Carlo simulation, adaptive sampling, etc. For example, to propagate the uncertainty from the inputs to the outputs, each time a (pseudo-)random realization is drawn from the probability distribution of the input parameters to calculate the corresponding output by running the simulation. After repeating this practice for a while, a probability distribution will be obtained for the outputs. Then, we can calculate the statistical moments, such as the min/max, empirical Probability Distribution Function (PDF)/Cumulative Distribution Function (CDF), based on the ensemble of calculations.

- Methods based on most popular point (MPP): first-order and second-order reliability methods.
- Methods based on functional expansion: Karhunen–Loeve exapansions (KLE), Neumann expansion.
- Methods based on local expansion: perturbation method, Taylor series, etc.
- Methods based on numerical integration: dimension reduction and full factorial numerical integration.

1.5.2 Inverse Uncertainty Management

Inverse uncertainty management deals with the minimization of the discrepancy between some measurements of a systems and simulated values of that system. This technique is called bias correction, or more specifically in reservoir engineering, history matching. The goal is to estimate the uncertain parameters of the system that is called parameter calibration. This technique is more complicated than the forward uncertainty management, since it is a model updating technique. The general methods of inverse uncertainty management (history matching) are:

- **Bias correction**

  This approach quantifies the difference between the measured and simulated output of the model. Equation 1.2 shows the general formulation for bias correction:

  $$y(x) = y^m(x) + \delta(x) + \varepsilon, \qquad (1.2)$$

  in which $y(x)$ and $y^m(x)$ are the simulated and measured (observed) outputs of the system with respect to inputs x, respectively. $\delta(x)$ is the difference between the simulated and measured values, and ε is the experimental measurement error. The ultimate goal is to minimize $\delta(x)$.

- **Parameter calibration**

  In this approach, the goal is to estimate the unknown parameters of the model. Equation 1.3 is the general formulation used in this approach:

  $$y(x) = y^m(x, \theta^*) + \varepsilon, \qquad (1.3)$$

in which $y^m(x, \theta)$ is the output of the computer model with parameters θ, θ^* is the true values of the parameters. In this approach, it is tried to find θ or a probability distribution that well represents θ^*.

- **Bias correction and parameter calibration**

This approach combines the above methods together:

$$y(x) = y^m(x, \theta^*) + \delta(x) + \varepsilon. \tag{1.4}$$

This formulation provides the most comprehensive approach to update the model parameters with various sources of uncertainty. However, its computational burden is higher than other two approaches.

The most commonly used methods for inverse uncertainty management are the Bayesian approaches [23–26]. These methods can be applied to solving problems of the third and the most complicated inverse uncertainty management problem, bias correction and parameter calibration. There are two types of Bayesian approaches: (1) Modular Bayesian approach, (2) Fully Bayesian approach.

Modular Bayesian approaches

This approach is consisted of four modules. In addition to the observed data, it is needed to assume a prior distribution for the uncertain parameters in this approach. The four modules are:

1. Module 1: replace the computer code, $\hat{y}$, with a Gaussian process (GP) simulator in which the parameters θ are deterministically estimated using the computational results. The estimation process can be found in Appendix B of [27].

$$y^m(x, \theta) \sim GP\left((h^m)^T \beta^m, \sigma_m^2 R^m\left((x, \theta), (x', \theta')\right)\right), \tag{1.5}$$

where

$$R^m\left((x, \theta), (x', \theta')\right) = \exp\left\{-\sum_{i=1}^{d} w_i^m\left(x_i - x_i'\right)^2\right\} \exp\left\{-\sum_{i=1}^{r} w_{d+i}^m\left(\theta_i - \theta_i'\right)^2\right\}. \tag{1.6}$$

Dimension of the input is denoted by d, and dimension of the output is denoted by $r \cdot h^m$ represents the hyperparameters of the Gaussian Process, $\{\beta^m, \sigma_m, w_i^m, i = 1, \ldots, d + r\}$. These hyperparameters can be estimated by the maximum likelihood estimation.

2. Module 2: use another GP simulator to replace the error function between the measured and simulated values.

$$\delta(x) \sim GP\left(h^\delta(.)^T \beta^\delta, \sigma_\delta^2 R^\delta(x, x')\right), \tag{1.7}$$

where $\delta(x)$ is the error function, and

$$R^\delta(x, x') = \exp\left\{-\sum_{i=1}^{d} w_i^\delta\left(x_i - x_i'\right)^2\right\}. \tag{1.8}$$

Using the prior distribution and the data from measurements and simulations, the hyperparameters $\{\beta^\delta, \sigma_\delta, w_i^\delta, i = 1, \ldots, d\}$ can be estimated using the maximum likelihood estimation.

3. Module 3: After comprehending the GP's hyperparameters from module 1 and 2, the posterior distribution of the parameters can be estimated. This is done through the Bayes' theorem (more details in Sect. 2.3.1):

$$p(\theta|data, \varphi) \propto p(data|\theta, \varphi)p(\theta), \tag{1.9}$$

where φ indicates the fixed hyperparameters in modules 1 and 2.

4. Module 4: In this module, simulated values and the discrepancy function between the simulated and measured values are calculated.

Fully Bayesian approach

In the fully Bayesian approach, we need the prior distributions for both the unknown parameters, θ, and the hyperparameters, φ. This approach consists of three steps:

- Derivation the posterior distribution, $p(\theta, \varphi|data)$.
- Calculating $p(\theta|data)$ by marginalizing the φ out. Model calibration (history matching) is done through this step.
- Estimating the output and $\delta(x)$.

One of the shortcomings of this approach is the intractability of $p(\theta, \varphi|data)$ that results in problematic integrations. It is common to use Markov Chain Monte Carlo (MCMC) methods to approximate these integrations [28–30]. MCMC is one of the powerful techniques to solve intractable inverse problems. Some of the MCMC methods are Metropolis sampling, Gibbs sampling, and Contrastive divergence. Interested readers are referred to [1] for more information on these methods.

One of the most challenging uncertainties in reservoir management in the geological uncertainty. The next chapter is dedicated to give more details on this kind of uncertainty, introducing types of geological uncertainty, explaining data used to quantify the geological uncertainty and how to parametrize it, and methods of generating realizations of the geologically uncertain parameters to quantify the range of uncertainty.

References

1. Goodfellow I, Bengio Y, Courville A (2016) Deep learning. The MIT Press
2. Miller DC, Ng B, Eslick J, Tong C, Chen Y (2014) Advanced computational tools for optimization and uncertainty quantification of carbon capture processes. Comp Aided Chem Eng 34:202–211. https://doi.org/10.1016/B978-0-444-63433-7.50021-3
3. Mirzaei-Paiaman A, Santos SMG, Schiozer DJ (2021) A review on closed-loop field development and management. J Petrol Sci Eng 201:108457. https://doi.org/10.1016/j.petrol.2021.108457
4. Santos SMG, Gaspar ATFS, Schiozer DJ (2021) Information, robustness, and flexibility to manage uncertainties in petroleum field development. J Petrol Sci Eng 196:107562. https://doi.org/10.1016/j.petrol.2020.107562
5. Shirangi MG (2019) Closed-loop field development with multipoint geostatistics and statistical performance assessment. J Comput Phys 390:249–264. https://doi.org/10.1016/J.JCP.2019.04.003
6. Hui Z, Yang L, Jun Y, Kai Z (2011) Theoretical research on reservoir closed-loop production management. Sci China Technol Sci 54(10):2815–2824. https://doi.org/10.1007/S11431-011-4465-2
7. Overbeek KM, Brouwer DR, Neavdal G, Kruijsdijk CPJW van, Jansen JD (2004) Closed-loop qaterflooding. Earth Doc. https://doi.org/10.3997/2214-4609-PDB.9.B033
8. Brouwer DR, Nævdal G, Jansen JD, Vefring EH, Van Kruijsdijk CPJW (2004) Improved reservoir management through optimal control and continuous model updating. Proceedings—SPE Annual Technical Conference and Exhibition. 26:1551–1561. https://doi.org/10.2118/90149-MS
9. Morosov AL, Schiozer DJ (2017) Field-development process revealing uncertainty-assessment pitfalls. SPE Reservoir Eval Eng 20(03):765–778. https://doi.org/10.2118/180094-PA
10. Hanea R, Evensen G, Hustoft L, Ek T, Chitu A, Wilschut F (2015) Reservoir management under geological uncertainty using fast model update. In: SPE Reservoir Simulation Symposium. Houston, TX
11. Mohammadi M, Ahmadi M, Kazemi A (2020) Comparative study of different risk measures for robust optimization of oil production under the market uncertainty: a regret-based insight. Comput Geosci 24(3):1409–1427. https://doi.org/10.1007/S10596-020-09960-7
12. Yeten B, Durlofsky LJ, Aziz K (2003) Optimization of nonconventional well type, location, and trajectory. SPE J 8(03):200–210. https://doi.org/10.2118/86880-PA
13. Mirzaei-Paiaman A, Sabbagh F, Ostadhassan M, Shafiei A, Rezaee R, Saboorian-Jooybari H, Chen Z (2019) A further verification of FZI* and PSRTI: newly developed petrophysical rock typing indices. J Petrol Sci Eng 175:693–705. https://doi.org/10.1016/J.PETROL.2019.01.014
14. Mirzaei-Paiaman A, Ostadhassan M, Rezaee R, Saboorian-Jooybari H, Chen Z (2018) A new approach in petrophysical rock typing. J Petrol Sci Eng 166:445–464. https://doi.org/10.1016/J.PETROL.2018.03.075
15. Mirzaei-Paiaman A, Saboorian-Jooybari H, Chen Z, Ostadhassan M (2019) New technique of True Effective Mobility (TEM-function) in dynamic rock typing: reduction of uncertainties in relative permeability data for reservoir simulation. J Petrol Sci Eng 179:210–227. https://doi.org/10.1016/J.PETROL.2019.04.044
16. Roggero F, Hu LY (1998) Gradual deformation of continuous geostatistical models for history matching. In: SPE Annual Technical Conference and Exhtition, pp 221–236. https://doi.org/10.2523/49004-ms
17. Gautier Y, Nœtinger B, Roggero F (2004) History matching using a streamline-based approach and gradual deformation. SPE J 9(01):88–101. https://doi.org/10.2118/87821-PA
18. Lange AG (2009) Assisted history matching for the characterization of fractured reservoirs. AAPG Bull 93(11):1609–1619. https://doi.org/10.1306/08040909050
19. Yousefzadeh R, Sharifi M, Rafiei Y, Ahmadi M (2021) Scenario reduction of realizations using fast marching method in robust well placement optimization of injectors. Nat Resour Res 30:2753–2775. https://doi.org/10.1007/s11053-021-09833-5

20. Yousefzadeh R, Ahmadi M, Kazemi A (2022) Toward investigating the application of reservoir opportunity index in facilitating well placement optimization under geological uncertainty. J Petrol Sci Eng 215:110709. https://doi.org/10.1016/J.PETROL.2022.110709

21. Canchumuni SWA, Emerick AA, Pacheco MAC (2018) History matching channelized facies models using ensemble smoother with a deep learning parameterization. In: 16th European Conference on the Mathematics of Oil Recovery, ECMOR 2018, September. https://doi.org/10.3997/2214-4609.201802277

22. Lee SH, Chen W (2009) A comparative study of uncertainty propagation methods for black-box-type problems. Struct Multidiscip Optim 37(3):239–253. https://doi.org/10.1007/s00158-008-0234-7

23. Santoso R, He X, Alsinan M, Kwak H, Hoteit H (2021) Bayesian long-short term memory for history matching in reservoir simulations. In: SPE Reservoir Simulation Conference. https://doi.org/10.2118/203976-MS

24. Eltahan E, Ganjdanesh R, Yu W, Sepehrnoori K, Drozd H, Ambrose R (2020). Assisted history matching using Bayesian inference: application to multi-well simulation of a Huff-n-Puff Pilot test in the Permian Basin. https://doi.org/10.15530/urtec-2020-2787

25. Elsheikh AH, Jackson MD, Laforce TC (2012) Bayesian reservoir history matching considering model and parameter uncertainties. Math Geosci 44(5):515–543. https://doi.org/10.1007/s11004-012-9397-2

26. Gardner P, Lord C, Barthorpe RJ (2020) Bayesian history matching for structural dynamics applications. Mech Syst Signal Process 143:106828. https://doi.org/10.1016/j.ymssp.2020.106828

27. Wu X, Kozlowski T, Meidani H, Shirvan K (2018) Inverse uncertainty quantification using the modular Bayesian approach based on Gaussian process, part 1: theory. Nucl Eng Des 335(June):339–355. https://doi.org/10.1016/j.nucengdes.2018.06.004

28. Liao Q, Zeng L, Chang H, Zhang D (2019) Efficient history matching using the markov-chain Monte Carlo method by means of the transformed adaptive stochastic collocation method. SPE J 24(4):1468–1489. https://doi.org/10.2118/194488-PA

29. Yustres Á, Asensio L, Alonso J, Navarro V (2012) A review of Markov Chain Monte Carlo and information theory tools for inverse problems in subsurface flow. Comput Geosci 16(1):1–20. https://doi.org/10.1007/s10596-011-9249-z

30. Liu B, Liang Y (2017) An introduction of Markov chain Monte Carlo method to geochemical inverse problems: reading melting parameters from REE abundances in abyssal peridotites. Geochim Cosmochim Acta 2017(203):216–234. https://doi.org/10.1016/j.gca.2016.12.040

Chapter 2
Geological Uncertainty Quantification

Abstract Our knowledge from underground reservoirs is not complete and is limited to some sparse core and log data, seismic data, geological interpretations, etc. This limited knowledge leads to a significant extend of uncertainty that is common in reservoir modeling and characterization. This kind of uncertainty is known as the geological uncertainty since the uncertainty is present in geological parameters, such as permeability, porosity, fluid contacts, reservoir compartmentalization, fault transmissibility, existence, type, and characteristics of aquifer, etc. Accordingly, geological uncertainty can have different scales including the macro- and micro-scale geological uncertainties. This chapter introduces the geological uncertainty, geological uncertainty scales, geological prior information, structural and stratigraphic uncertainties, geological parametrization, use of seismic and petrophysical data in uncertainty quantification, exploring the range of scenarios, geological realizations, and the geostatistical methods to generate geological realizations. The main geostatistical techniques for generating the realizations of the uncertain parameters, including the kriging-based, object-based, and multiple-point geostatistical methods, are introduced, and their applications, advantages and disadvantages are presented.

Keyword Geological uncertainty · Geological realizations · Geostatistics · Prior information · Bayesian rule · Structural uncertainty · Stratigraphic uncertainty · Multiple-point geostatistics · Geological parametrization

2.1 Geological Uncertainty Scales

Depending on the scale of the uncertain parameters, geological uncertainty can be divided into two categories: Macro-scale geological uncertainty and micro-scale geological uncertainty.

Macro-scale geological uncertainty refers to uncertainty in parameters in the field/reservoir scale. These parameters encompass reservoir compartmentalization, depth of water/oil and gas/oil contacts, existence of aquifers, aquifer characteristics, etc. The uncertainty in these parameters can be represented via a range of variations or a probability distribution and is easier to be accounted for in field development activities than micro-scale uncertainty. Reservoir compartmentalization can occur

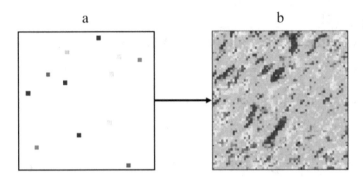

Fig. 2.1 a Measured (known) permeability values, **b** permeability distribution

due to the existence of a sealing fault. Since the existence of faults is of a great uncertainty, it leads to a macro-scale uncertainty. Also, existence of an aquifer and its characteristics is a macro-scale uncertainty since although aquifers include properties like permeability and porosity, they are used as average values and not in pore-scale quantities. Figure 1 in [1] demonstrates a compartmentalized reservoir.

Micro-scale uncertainty refers to uncertainty in those reservoir parameters that are in the scale of pores. Pore-scale parameters are grid properties such as permeability distribution, porosity distribution, net/gross ratio distribution, etc. The uncertainty in these parameters are more complex and computationally expensive to be accounted for. A reservoir model may contain hundreds of thousands to millions of grid blocks which their properties must be provided. However, due to our limited measurements of the grid properties only at particular locations, we cannot certainly infer their values all over the reservoir. For example, permeability is only measured at well locations through core samples. However, we have not a direct measurement of permeability at other locations. Consequently, this kind of parameters impose a large degree of uncertainty on the reservoir characterization problem. Figure 2.1 shows the measured data (Fig. 2.1a) and the permeability distribution (Fig. 2.1b) of a reservoir model. As can be seen, the number of known permeability values is extremely fewer than unknown values. Thereby, we rely on these limited measurements of the uncertain parameters to propagate the pore-scale properties throughout the entire reservoir model.

2.2 Stratigraphic and Structural Uncertainties

A conceptual model is the basis of an underground reservoir model, which includes the understanding of the modeler of the system, as well as some assumptions about a specific case [2]. A conceptual model is comprised of a geological model that combines chemical and physical phenomena (flow equations), discretization in time and space domains, and initial and boundary conditions. Technically, a geological

model is comprised of multiple structural elements, which are obtained from stratigraphic interpretations [3]. The main roots of uncertainty in predictions of a model
are the: (1) input data (seismic data, production data, hard data, etc.), (2) model characteristics data (e.g., porosity, permeability, etc.), (3) conceptual model, (4) modeling
constraints—environmental, economic, political [4, 5]. Most of studies in uncertainty
management and quantification focus on the uncertainty in model characteristics
[6–9] and input data [10, 11]. Nonetheless, in reservoir modeling, uncertainty also
exists in conceptualization and geological interpretation. This uncertainty is denoted
as conceptual geological uncertainty [3]. We can classify the conceptual geological
uncertainty into three main groups:

1. Stratigraphic uncertainty,
2. Geological Structural Uncertainty,
3. Conceptual Uncertainty.

Figure 1 in [3] shows the types of conceptual geological uncertainties and the
methods to quantify each type. These methods are explained in more details in the
following subsections. Note that for quantifying homogenous areas less data are
required, but more data are needed to quantify the uncertainty of heterogeneous
areas.

Stratigraphic uncertainty deals with uncertainties related to determining the location of the surfaces of stratigraphic layers, which describe the geological model.
On the other hand, geological structural uncertainty points to uncertainties related
to the combination of geological structural constituents, including faces and lenses.
All uncertainties on assumptions on the recognized geological structural elements,
2D or 3D flow, dominating flow regime (radial, linear, fractured flow), initial conditions, boundary condition, etc. are regarded as conceptual uncertainty. This means
that as we go from stratigraphic to conceptual uncertainty, uncertainty increases.
Conceptual uncertainty is dominant if there are less data and information and the
geology is heterogeneous, while when there are comprehensive data and the geology
is homogeneous, stratigraphic uncertainty is of a greater importance [3].

Each type of uncertainty is addressed by different methods (see Fig. 1 in [3]).
For example, an ensemble approach, which takes benefit from multiple conceptual
models, is used to tackle the conceptual uncertainty. The ensemble may contain
several geological models with varying structural definitions [12] and different
boundary conditions or model codes [13, 14].

Geological structural uncertainty is usually a main cause of uncertainty in reservoir
engineering with a large extent of uncertain parameters and less geological knowledge [15–18]. Geological structural uncertainty can be quantified by category-based
tools by simulating the combination of a set of predefined geological scenarios, such
as sand, limestone, and dolomite. Some well-proven methods to generate geological structures in a stochastic manner include the TProGS [19, 20], multiple-point
geostatistics [21–24], Bayesian frameworks [25], and methods based on Markov
chains [26, 27]. Thore et al. [28] introduced a systematic approach to determination and management of structural uncertainties based on the probability field technique modified for triangulated surfaces handling the correlations between horizons

of faults. They also provide some applications that needed quantifying structural uncertainties, including the rock volume distribution, well trajectory optimization, and using structural uncertainty in history matching as the uncertain parameter. Their approach consisted of three main steps including the [28]:

1. Building a structural model

   a. Selection of main sources of uncertainty,
   b. Assessment of each source of uncertainty (amplitude, direction & correlation length),
   c. Setting constraints (fault-to-fault and horizon-to-fault).

2. Computing Realizations

   a. Move horizons according to uncertainty using vertical correlation,
   b. Move faults according to horizons,
   c. Move faults according to uncertainty and fault-to-fault constraints,
   d. Stick horizons to faults.

3. Post processing

   a. Check conformity of the realizations,
   b. Rock volume distribution,
   c. Well trajectory optimization,
   d. Spill point determination.

Stratigraphic uncertainty is quantified by boundary-based tools by which the location of the layout of geological structures is analyzed, for example, the no flow boundary. Boundary-based methods are generally based on kriging interpolation [29, 30], in which the kriging variance is used to estimate the uncertainty of the geological surfaces. One of the limitations of these methods is that they do not take into account the data uncertainties in interpreting the borehole data. Another limitation is that the quantifications are smooth because of using smooth surfaces due to utilizing the kriging variance ignores the local and small-scale variations [29]. Gong et al. [31] also pointed out the limited number of investigations on quantifying stratigraphic uncertainty and proposed a stochastic Markov random field-based technique to tackle stratigraphic uncertainty in geotechnical design studies. The studies showed that it is better to use a probabilistic approach rather than a simplified deterministic method to account to the stratigraphic uncertainty.

2.3 Geological Prior Information

To resolve the geological uncertainty, geological information is needed. This information can include information on the structure, lithology, stratigraphy, pore volume, permeability, porosity, etc. In this regard, some theoretical and applied problems are solved by transferring specific geological information to other domains. For example:

- information about geological structures in a specific environment can be used to estimate underground properties at locations away from observation wells to estimate hydrocarbon reserves [32],
- evaluating geo-hazard risks can be used for valuing of life and insurance of properties [33],
- combining the correlation between sea-level and growth of corals with the distribution of old reef corals can be used to predict both fluctuations in past sea-level, and also distribution of new reefs with respect to future sea-level alterations [34, 35],
- estimating tectonic stability in a region can be used to store toxic waste in underground formations with a minimized risk of leakage [36].

In most cases, this geological information is provided a priori, which means the information given before any new observation is obtained from the underground formation and is used to find the solution. Such information is known as *geological prior information* [37].

It should be pointed out that not all prior information is useful and we should use only reliable geological prior information. In this regard, the following requirements have to be fulfilled:

(a) Consulting several geological experts and reconciling the conflicting point of views,
(b) The confidence or uncertainty of all prior information is determined; otherwise, their reliability and value cannot be known,
(c) Gathering more information is more quantitative. Also, clearly defining the qualitative information or assumptions help to evaluate the uncertainty or risk in the final result.

2.3.1 How to Use and Analyze Geological Prior Information

Consider $P(Q)$ the probability of an event or hypothesis (Q) to be true if the event space is discrete. If the event space is continuous, $P(Q)$ will be the probability density function (PDF) of Q. Another notation is the conditional probability noted as $P(Q|R)$, which means the probability of occurring Q given that R has already happened (or is true). Therefore, the Bayes' Rule can be used to calculate the conditional probability $P(Q|R)$ as follows:

$$P(Q|R) = \frac{P(R|Q)P(Q)}{P(R)}, \qquad (2.1)$$

Here, $P(Q)$ is the prior probability of event Q, $P(R|Q)$ is the likelihood of event R given Q, and $P(R)$ is the probability of event R.

Now, assume that we are going to answer a geoscientific question which the available geological information may affect it. Also, say that the question is raised in such a way that it is needed to constrain a model vector m to answer that question. Model m can be the distribution of permeability throughout a hydrocarbon reservoir. In this regard, related background information might consist: geological information acquired from neighboring reservoirs, general information about the regional geology, diagenetic, tectonic or climatic history, expected data uncertainties, and all known methods to address different problems [37]. All mentioned background information is denoted by symbol B. We can assume that B consists of geological G and non-geological G' knowledge, and moreover, the geological knowledge includes quantitative G_q and non-quantitative $G_{q'}$ parts. Therefore, we can represent $B = \left[G_q, G_{q'}, G' \right]$.

By acquiring new data d, we can impose more constraints on model vector m. By using the Bayes' Rule, we can take benefit from both new data and prior information B to further constrain m. Since determining the individual effect of d and B on constraining m is not an easy task, using the Bayes' Rule is beneficial to tackle this issue. As there is uncertainty in acquired and prior information, we can never reach certainty in our knowledge about m. Therefore, it is better to use a probability distribution to represent the information on m. To do so, m, d, and B are substituted to Eq. 2.1:

$$P(m|d, B) = \frac{P(d|m, B)}{P(d|B)} P(m|B), \tag{2.2}$$

where $P(m|d, B)$ is the *posterior distribution*, which is obtained by incorporating the prior knowledge B and new data d to the model vector m. The posterior distribution can be computed for any m by multiplying the two term on the right hand side (RHS) of Eq. 2.2. The first term in the RHS of Eq. 2.2 is the relative likelihood term, calculated as the ratio of the probability of observing d given model m (numerator) to the probability of data when m is not available (denominator). Finally, the second term in the RHS of Eq. 2.2 is the *prior distribution* of model vector m, and is our prior knowledge of m before acquiring any data d.

To cast more light on the application of the prior knowledge in characterizing a reservoir, let us consider the following example. In this example, we are going to constrain the lithology of a reservoir rock. Let m represent different possible lithologies. Therefore, m can have one of the following values: Sandstone (SS), Carbonate (C), Dolomite (D), or Shale (S). Related prior information B may include: regional geological and stratigraphic history, geological information from neighboring reservoirs, and sedimentary environment. Consequently, $P(m|B)$ will contain all possible probabilities on the reservoir lithology based on only the above prior information. Now, assume that some core samples are obtained from the reservoir to provide a dataset d. The numerator in the first term of the RHS of Eq. 2.2, $P(d|m, B)$, represents the probability of recording dataset d if one of the possible values of m was true (e.g., the lithology was sandstone) with the given prior information B. The denominator $P(d|B)$ is the probability of observing d without any information about the true lithology of the reservoir, which plays the role of a normalization term to ensure that the posterior distribution has a unity integral.

To have a numerical example, assume that based on the prior information B, we know that the reservoir has deposited in deep waters. Then, we can assign a low prior probability to SS to be the lithology of the reservoir (i.e., $P(m = SS|B)$). However, the probability of other lithologies are considered equal. Thereby, the following prior distribution can be defined:

$$P(m = SS|B) = \frac{1}{10}, and\, P(m = C|B) = P(m = D|B) = P(m = S|B) = \frac{3}{10}.$$

In addition, the probability of observing data d given each of possible values for m is:

$$P(d|m = S, B) = P(d|m = SS, B) = P(d|m = D, B) = \frac{1}{5}\, and\, P(d|m = C, B) = \frac{2}{5}.$$

This means that the probability of observing d if the lithology is Carbonate is twice that of all other lithologies.

To this end, the likelihood in the numerator and the prior probability term on the RHS of Eq. 2.2 are defined. Now, we can treat the other two terms. We must assume a value for the denominator on the RHS in such a way that the sum of posterior probabilities for all values of m is one. For this example, we can calculate the denominator by marginalizing out the conditional probabilities of $P(d|m, B)$ for all values of m.:

$$
\begin{aligned}
P(d|B) &= P(d|m = SS, B)P(m = SS|B) \\
&+ P(d|m = S, B)P(m = S|B) \\
&+ P(d|m = C, B)P(m = C|B) \\
&+ P(d|m = D, B)P(m = D|B) \\
&= \frac{1}{5}\cdot\frac{1}{10} + \frac{1}{5}\cdot\frac{3}{10} + \frac{2}{5}\cdot\frac{3}{10} + \frac{1}{5}\cdot\frac{3}{10} = \frac{14}{50}.
\end{aligned}
$$

Now, the posterior distribution can be calculated:

$$P(m = [SS, S, C, D]|d, B) = \left[\frac{1}{14}, \frac{3}{14}, \frac{6}{14}, \frac{3}{14}\right],$$

where the values in the square brackets are the probabilities of each lithologies in m as listed in the square brackets on the left hand side. According to the posterior distribution, the probability of carbonate, C, is twice the probability of dolomite and shale, as could be informed based on the likelihood distribution (data information). Nonetheless, after incorporating the *prior information*, the probability of carbonate, C, was shown to be three times the probability of sandstone, SS. Although this was a simple example, it can be generalized to more sophisticated problems such as inferring the posterior permeability distribution by incorporating the prior knowledge and likelihood of the observed production/4D seismic data.

2.4 Use of Seismic and Petrophysical Data in Uncertainty Quantification

It is mathematically and statistically challenging to model the spatial distribution of petrophysical properties in a hydrocarbon reservoir because of sparse direct measurements of these properties [38, 39]. Since the direct measurements of petrophysical data are rare, it is required to use indirect measurements as conditioning data to be able to model the distribution of spatial petrophysical data throughout the reservoir. In uncertainty quantification terminology, direct measurements of the property of interest are known as hard data, and indirect measurements are known as soft data. Typically, hard data are rare, but soft data are abundant. For example, a co-kriging can use porosity at well locations as hard data (petrophysical data) and acoustic impedance (as soft data) from seismic interpretations, which is a seismic attribute, to estimate the porosity at other locations [40]. Seismic attributes provide a link between petrophysical properties and seismic data of the reservoir, which are directly or indirectly related to rock properties of the field [41]. In addition, seismic attributes can be used to find correlations between wells. Seismic facies data are another example of soft data used for quantifying facies uncertainties. A limitation of seismic data is their low resolution with low signal to noise ratio. Multiple approaches based on Bayesian inversion theory [42, 43], geostatistics [44–46], and machine learning [47] have been used to infer the posterior facies distribution given a set of seismic measurements. Note that, there is not a unique solution to problem since the same seismic response can be produced by various spatial distribution of facies sequences. Thereby, it is needed to generate a probability distribution of the facies resulting in the observed seismic measurements. Bayesian approaches can be applied to various geophysical inversions, such as seismic inversion, facies classification, and petrophysical inversion [48–50]. From a Bayesian point of view, the result of an inverse problem is a probability distribution of the property of interest for a set of model hyperparameters [51]. If the property of interest is discrete, e.g., facies classification, the uncertainty is quantified by the posterior probability.

Given a simple problem, in which, there is no spatial correlation, the Bayesian approach is relatively simple to implement [52]. Having X as a measured property (at a specific location), we use the conditional probability $P(\pi = i|X)$ to represent the probability that the facies π belongs to class $i \in \Omega = \{0, \ldots, K - 1\}$ (where K is the total number of classes). In real applications, K is usually less than 10. Based on the Bayes' Rule (Eq. 2.2), $P(\pi = i|X)$ is calculated as the product of the likelihood of observing X given the facies i ($P(X|\pi = i)$) times the prior knowledge of the facies i denoted as $P(\pi = i)$ divided by the marginal probability of the conditional probability $P(X|\pi = i)$ for all facies classes ($P(X)$):

$$P(\pi = i|X) = \frac{P(\pi = i)P(X|\pi = i)}{\sum_j P(\pi = j)P(X|\pi = j)}.$$

If X is multivariate, the above formulation can be generalized to multiple measurements at the same spatial location. Nonetheless, we cannot assume spatial independence for the facies distribution in underground formations. There is spatial correlation in the distribution of facies since the deposition is geologically continuous and stacked stratigraphically. Observed seismic data introduces augmented spatial correlation. Therefore, seismic data is used as conditioning (soft) data for uncertainty quantification.

Geostatistical inversion tools presume that some prior information is known and stationary (e.g., semi-variograms, training images, etc.); however, this assumption is not true and results in underestimating the uncertainty of the model. Petrophysical data, such as porosity and permeability, are used to derive the empirical semi-variograms that can be used in kriging-based quantification methods or can be used as hard data in multiple-point statistics to find the data events in the training image.

2.5 Exploring the Range of Scenarios

Exploring the range of scenarios involves sampling from the probability distribution of the parametrized model properties using the Monte Carlo simulation [53]. Monte Carlo methods are more suitable for numerical problems than other conventional numerical approaches. Nevertheless, it is required to randomly sample from probability distributions of high dimensional parameters to be able to use Monte Carlo methods, which is a complex and time consuming task. Also, most of the time we do not have access to the probability distribution of the parameters. Generally, sampling from such distributions or estimating expectations with respect to them are done using the following methods proposed by Hastings [54]:

1. Factorizing the high dimensional distribution into the product of one-dimensional conditional distributions from which samples may be obtained.
2. Using importance sampling, which can also be used to reduce the variance. Therefore, to compute the expectation of $f(x)$, $(E_p(f))$:

$$J = \int f(x)p(x)dx = E_p(f),$$

instead of drawing independent samples $x_1, \ldots, x_N$ from the probability density function, $p(x)$, and using the approximation $\hat{J}_1 = \Sigma f(x_i)/N$, we can substitute $p(x)$ by another probability density function, $q(x)$, to sample from and use the approximation $\hat{J}_1 = \Sigma\{f(x_i)p(x_i)\}/\{q(x_i)/N\}$. If $q(x)$ is simpler than $p(x)$ and easier to sample from, it resolves the issues related to directly sampling from $p(x)$ [54, 55]. However, if the number of dimensions is large, it will be a difficult approach to use as the weights $w(x_i) = p(x_i)/q(x_i)$ maybe extremely small for reasonable values of N, or some of them may be too large.

3. Using a simulation method: this is useful when sampling from $p(x)$ is difficult or impossible or $p(x)$ is unknown. Consequently, instead of sampling from $p(x)$, we sample from another distribution $q(y)$ and infer the sample x values based on the corresponding y values. If the goal is to sample from the conditional distribution of $x = g(y)$, given $h(y) = h_0$, it is not recommended to use the simulation method if $P(h(y) = h_0)$ is small due to the low probability of $h(y) = h_0$ although the large size of the sample from $q(y)$.

Monte Carlo methods using Markov chains (MCMC) provide a convenient way to sample from complex and high dimensional distributions [53, 54, 56]. The primary characteristics of MCMC techniques for drawing samples from a distribution with a density $p(x)$ are:

i. The only dependence of the calculations to $p(x)$ is via proportions of the form $p(x'/x)$, where x' and x are samples. As a consequence, there is no need to know the normalizing constant in the denominator of Eq. 2.2 and factorize $p(x)$. Additionally, there is no need to treat the conditional distributions in special ways since the methods make it easy to sample from conditional distributions.

ii. Each time a Markov chain is simulated, a sequence of samples is generated. This means that the obtained samples are correlated and more care is required in computing the standard deviation of an estimate and analyzing the error of an estimate compared to the case where samples are independent.

Importance sampling [54, 57], Metropolis sampling [56], Metropolis-Hasting sampling [58, 59], Gibbs sampling [55, 60], Hamiltonian Monte Carlo sampling [61], and contrastive divergence [62–64] are some of the most used MCMC methods to explore the range of scenarios in uncertainty quantification problems. Each time the simulation is repeated, a new sample/scenario is generated. It is common to repeat Markov chain techniques to stochastically update the state of parameters until it begins to produce samples from the equilibrium distribution, which is known as burning the Markov chain [55]. At this point, we can ensure that the range of scenarios is explored. As stated above, the samples obtained from the Markov chain are correlated. To obtain truly independent samples, one can run multiple Markov chains in parallel.

2.6   Geological Parametrization

A better posterior distribution is a distribution with the maximum probability. Therefore, according to Eq. 2.2, the likelihood of observing new data, d, should be maximized since the prior probability and data event probability cannot be changed. In most cases, calculating $P(d|B)$ is intractable [55]. Consequently, by ignoring $P(d|B)$, we can rewrite Eq. 2.2 as an approximation:

$$P(m|d, B) \approx P(d|m, B)P(m|B). \tag{2.4}$$

Equation 2.4 is known as the *maximum* a posteriori (MAP). Accordingly, to maximize the posterior probability, the likelihood of observing the data has to be maximized. To do so, we have to adjust the model vector m in such a way to minimize the discrepancy between the simulated and actual observed data, which is known as data assimilation or history matching. If vector m has a large number of elements (i.e., there are lots of uncertain parameters), updating m will be cumbersome. Consequently, it is required to reparametrize the vector m so that the number of parameters involved in history matching is reduced. This step is known as geological parametrization in which the original uncertain parameters are reparametrized so that they can be represented by fewer number of parameters [65, 66]. Usually, this parametrization results in dimensionality reduction of the problem so that updating the model vector becomes easier and helps preserving the geologic realism of the true model. For example, instead of updating a huge number of original petrophysical data, fewer parameters are updated. These parameters can include pilot points in which some points are selected to be updated to run a kriging regression thereafter to calculate the value of the property of interest at other locations [67, 68]. Another approach is to update a Gaussian noise instead of original data and update the model parameters by incorporating the generated noise to the current model parameters. This technique is known as gradual deformation [69]. There are other parametrization techniques that deal with mapping the original model parameters into another space with lower dimensions. These techniques include linear and non-linear methods which are fully covered in Chap. 4.

2.7 Geological Realizations

The scarce hard data shown is Fig. 2.1 are used to generate the permeability distribution throughout the reservoir by geostatistical tools. Since the methods used for generating the distribution take benefit from random processes, each time we repeat the generation process, a new distribution is generated. These distributions are known as realizations of the uncertain parameters. Totally, a realization of the uncertain parameters can be generated by sampling from their probability distribution. Since permeability is a random function, each time we generate a permeability distribution, in fact, we sample from the probability distribution of permeability at different locations. Realizations can be generated for other kind of geological parameters such as the depth of fluid contacts, fault transmissibility, net-to-gross ratio, etc. For example, if the depth of gas-oil contact (GOC) follows a uniform distribution of $U[4500, 5000]$, and the fault transmissibility follows a uniform distribution of $U[0, 1]$, then multiple realizations of GOC and fault transmissibility can be generated as shown in Table 2.1. In this example, five samples are drawn from the distribution of the fault transmissibility and GOC that produces five geological realizations.

Although sampling from the probability distribution of macro-scale uncertain parameters is not too complicated, it is very challenging to sample from the probability distribution of micro-scale uncertain parameters such as permeability and

Table 2.1 Five geologic
realizations of fault
transmissibility and gas-oil
contact depth

# of realization	Fault transmissibility	GOC (ft)
1	0.00	4600
2	0.20	4580
3	0.50	5000
4	0.62	4803
5	0.90	4620

porosity. This is because we have very scarce number of directly measured values of
these parameters; therefore, we don't have access to their probability distribution at
all locations. Consequently, different geostatistical methods have been developed
over years to generate plausible geological realizations of the micro-scale reservoir
properties, such as the permeability, porosity, and net-to-gross ratio. In the following,
some of the most common realization generation methods are described.

2.8   Geostatistical Methods of Generating Geological Realizations

Geostatistics is different from conventional statistics in three ways [70]: (1) conventional statistics utilizes traditional mathematics; however, geostatistics incorporates
geology into the conventional statistics; (2) we can apply conventional statistics
to a wide range of problems including geology; however, geostatistics is limited
to geology; (3) geostatistics is far better than conventional statistics in terms of
randomicity and structure of the space of the parameters since geostatistics is able to
use furthest spatial information to locally and globally interpolate and obtain accurate
results.

Geostatistical methods of generating geological realizations can be classified into
four main categories:

1. Kriging-based methods (two-point statistics),
2. Object-based methods,
3. Pixel-based methods,
4. Multiple-Point Geostatistics (MPS).

These methods are discussed in the following subsections.

2.8.1   Kriging-based Methods

Kriging is a Gaussian process and is a kind of regression method that interpolates
between spatially distributed points [71]. Kriging is different from linear regression
in two ways: (a) unlike linear regression, it is not necessary for the variables to

be independent; (b) in kriging, non-random sampling replaces random sampling in conventional regression. Kriging has its root from mining geology; however, it has become one of the important tools in geostatistics. One of the advantages of kriging is that it not only generates an estimation of the parameter understudy at a specific location, but also gives an error of estimation which is a measure of uncertainty for that estimation. Kriging takes benefit from the variogram of the quantity of interest to conduct the regression process. Variogram is a measure that indicates the variability of a geostatistical parameter, such as the permeability, over a spatial region introducing a geological knowledge to the regression process. Interested reader can refer to [72] for more details on variogram and its types.

Kriging takes benefit from empirical observations of the parameter understudy, which are known as hard data, which is unlike other regression methods such as linear regression that presumes a particular model for regression. The hard data are used to derive the variogram. The basic mathematical equations of kriging were developed by Daniel Krige and further advanced by Matheron [73]. The general formulation of kriging at a specific location, x_0, is given in Eq. 2.5.

$$Z^*(x_0) = \sum_{i=1}^{n} \lambda_i Z(x_i), \tag{2.5}$$

where $Z^*(x_0)$ is the target value at location x_0, λ_i is the weight of the ith known data, $Z(x_i)$ is the known value at location x_i, and n is the total number of known data. The ultimate goal of the kriging techniques is to find the appropriate values of λ_i. In this method, nearby points get more weight in the interpolation process than farther points.

Kriging can be classified into the following categories: (1) Simple kriging, (2) Ordinary kriging, 3- Universal kriging. The basics of the statistics and linear algebra behind these methods can be found in [74]. The core assumption of these methods is to have a constant mean over the kriging domain, which means that data must be stationary. The main drawback of kriging is that it relies on two-point statistics (distance); therefore, it produces smooth results [75]. This is because kriging cannot preserve the covariance between the kriged points. Stochastic simulation based on kriging methods can mitigate this issue [76]. Now, we are going to see their applications in generation of geological realizations.

Kriging-based methods of generating micro-scale geological realizations include the Stochastic Gaussian Simulation (SGSIM) [77], Truncated Gaussian Simulation (TGSIM) [46, 78], and Sequential Indicator Simulation (SISIM) [79]. SGSIM is one of the most used kriging-based methods for generating realizations of continuous petrophysical properties. The core calculations of SGSIM is one of the aforementioned kriging methods. However, the difference between SGSIM and kriging is that SGSIM uses previously kriged (simulated) points as known data for the next steps. This is done to preserve the covariance between the kriged points by proceeding sequentially which helps SGSIM to better model the heterogeneity of the reservoir.

Figure 30.1 in [80] illustrates the difference between the results of the SGSIM and kriging. The overall steps of SGSIM using the simple kriging are as follows [81]:

1. Use normal score transformation and transform the distribution of hard data into the normal-standard distribution. To transform an empirical distribution to the normal-standard distribution, the empirical cumulative distribution function (CDF) of the data is generated. Then, for each value of the experimental data, its corresponding CDF is found. Next, the corresponding value in the normal-standard distribution with a CDF equal to the CDF of the experimental data is chosen. This value is known as the normal score. This process is repeat for all other experimental data until their normal scores are found.
2. Use the transformed data and fit an empirical variogram to them to determine its range and sill.
3. Generate a random path of unsampled locations.
4. For each point along the random path, repeat the following steps:

   a. Search for all neighboring hard data and previously simulated points.
   b. Perform the Kriging at location $\boldsymbol{u}$ using the neighboring points to obtain the kriged estimates, $Y^*(\boldsymbol{u}) = \sum_{i=1}^{n} \lambda_i . Y(\boldsymbol{u}_i)$, where $Y(\boldsymbol{u}_i)$ is the value of the property of interest at location $\boldsymbol{u}_i$.
   c. Calculate the variance of kriged estimates: $Var\{Y^*(\boldsymbol{u})\} = C(0) - \sigma_{SK}^2(\boldsymbol{u})$, where $C(0)$ is the variance of the kriged estimates at $\boldsymbol{u} = 0$, and $\sigma_{SK}^2(\boldsymbol{u})$ is the variance of the simple kriging for estimate at location $\boldsymbol{u}$. This means that the variance of kriged estimates is lower at locations far from the hard data locations. This is known as the missing variance of kriging.
   d. Draw a random residual, $R(\boldsymbol{u})$, that follows a normal distribution with a zero mean and variance of the missing variance. Add this variance to the value obtained from kriging to get the simulated value: $Y_s(\boldsymbol{u}) = Y^*(\boldsymbol{u}) + R(\boldsymbol{u})$. This is equivalent to sampling a value from a normal distribution with kriging estimation and variance as its mean and variance, respectively.
   e. Add the simulated value to the set of known data to ensure that the covariance between the simulated values follows the fitted variogram model. This is the key idea of sequential simulation, that is, to consider previously simulated values as hard data so that we reproduce the covariance between all of the simulated values.
   f. Visit all locations in a random order. There is no theoretical requirement for a random order or path; however, practice has shown that a regular path can induce artifacts.

5. Transform the simulated and hard data to their original space to complete the generation of a realization of the uncertain parameter.
6. Repeat all steps with a new random path to generate multiple realizations. A different seed leads to a different sequence of random numbers and, as a consequence, a different random path and different residuals for each simulated node. Each realization is "equally likely to be drawn" and often called equiprobable.

According to above explanations, SGSIM seems to be computationally demanding, especially for 2D and 3D models. This computational burden becomes more important if new data are going to be added and the model should be updated. Therefore, all steps must be repeated [81]. Another shortcoming of SGSIM is the smoothness of the resulting realizations. Since SGSIM uses kriging to estimate the unsampled locations, the generated values are smooth and close to the mean. Also, since the primary assumption of kriging-based methods is the stationarity of the property of interest, it cannot simulate non-stationary distributions.

The primary idea of SISIM and TGSIM is based on SGSIM. The difference between the SISIM and SGSIM is that SISIM is used for simulating categorical data such as facies modeling. TGSIM is also used for simulating categorical data by truncating the simulated values by SGSIM into categorical values. Since simulated values are in a sequential order, by taking another path, a new realization can be generated. Interested readers can refer to [79] and [78] for more details on SISIM and TGSIM, respectively.

2.8.2 Object-based Methods

Object-based geostatistical simulation methods are among the most commonly used methods for generating facies realizations. The primary idea behind these methods is to distribute some objects throughout the model [82, 83]. A parametrization of the shape of the objects, their density and properties can be used to consider the geological uncertainty. Although this method is simply implemented, a major short-coming of this approach is its limited flexibility. This is because the objects are often simple analytical shapes. For example, channels are defined with sinusoids, lenses are defined with 3D ellipsoids, fractures and defined using 2D ellipsoids, and other regions are considered the background sand/shale (see Fig. 2.2). Therefore, they cannot capture the complex structures of underground geobodies. Consequently, these methods lack the ability to the preserve the geologic realism [84].

2.8.3 Multiple-Point Geostatistics (MPS)

As discussed above, one of the drawbacks of kriging-based (two-point) geostatistical methods is that they are unable to capture the heterogeneous and complex structures and yield smooth results. Multiple-Point Geostatistics (MPS) is introduced to over-come the bottlenecks of two-point geostatistics [21–23]. Another important aspect to be considered is the scarcity of hard data from the reservoir that asks for the application of multiple-point statistics (MPS) alongside kriging-based methods. Another advantage of MPS over two-point geostatistics is that the MPS methods do not have the prerequisite of the input to have a Gaussian distribution as most of the natural phenomena do not follow the Gaussian distribution. These methods are based on a

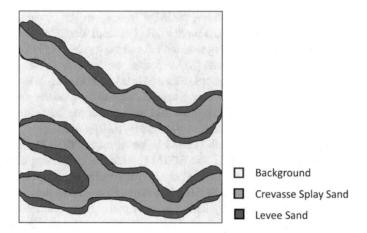

Fig. 2.2 Plan view of a conceptual model for fluvial facies: background of floodplain shales, levee border sands, and crevasse splay sands

Training Image (TI) and do not use two-point statistics. As a result, they are able to reproduce complex and heterogeneous structure of subsurface geological models.

Training images are some images representing the spatial structure of a subsurface model, which are created using prior information, such as seismic, core, log, petrophysical, and geophysical data. Multiple-point statistics can be extracted using these TIs which enable us to model complicated structures [85]. The role of this property becomes more important when the spatial connectivity is of a great importance in model applications [86–88]. However, one of the challenges in MPS methods is to provide a representative training image. Figure 30.2 in [89] illustrates some examples of training images.

There are three main techniques to create training images:

- From *Outcrop data*: An image of an outcrop can be a training image. Outcrops are primary sources of information of the geological structures [90]. In addition, the geometry and spatial continuity of the existing structures can be illustrated in 2D using the outcrop images for visual inspection.
- *Object-based methods*: Another way for creating structured models with categorical parameters is the objected-based methods [82, 83, 91]. Object-based methods are one of the most used methods to generate training images of facies models. An important part of any object-based modeling program is the geometric form and parameters used to represent each facies unit [92].
- *Process-based Methods*: These methods take benefit from physical processes of the porous media to create three-dimensional models [93–95]. The main bottleneck of these methods is that they are computationally demanding and need significant calibrations. Another limitation of these methods is that they are not general since they use the physical processes that are unique for each type of formation.

MPS methods perform on simulation grids, G, consisted of sampled and unsampled structured cells. The unsampled cells are estimated along a path that is either random or structured. This path is also known as the Simulation Path [89]. One of the most used simulation paths in sequential simulation algorithms is the Random path. This path is generated by assigning random numbers, with a different random seed for each realization, to the unsampled cells. Structured paths are known as Raster path, which are created based on a structural one-dimensional paths.

The unsampled points on the simulation grid are visited and estimated according to the simulation path. The estimation process is performed by finding a data event consisting of the neighboring sampled and simulated points. Data event consists of several known points. These known values can comprise both hard (conditioning) data and simulated cells. This data event is searched for by a template. A template consists of several points with a systematic organization that are used to find similar patterns in the training image [89]. After finding the similar pattern to the data event in the simulation grid, according to the MPS method used, the value is copied from the training image to the unknown location in the simulation grid (see Fig. 2.3). This process is repeated until all points are visited.

MPS methods can be divided into two main groups: (1) pixel-based, and (2) pattern-based methods. Pixel-based methods have the advantage of matching the well data; however, they cannot produce realistic realizations when dealing with complex models. Contrary to pixel-based methods, pattern-based methods are more accurate at modeling complex subsurface structures, but they don't exactly match the hard data. In pattern-based methods, a group of unknown points is simulated simultaneously. The major drawback of this kind of MPS method is their poor conditioning to hard data. This postulates the combination of pixel-based and pattern-based methods that leads to a hybrid framework. Unlike kriging-based methods, MPS methods do not have the assumption of stationarity and can be used for modeling non-stationary

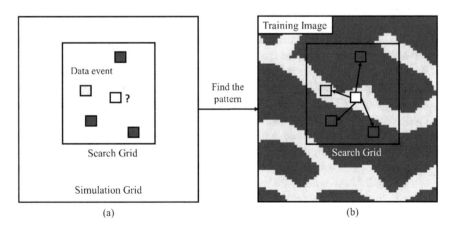

Fig. 2.3 Schematic of the Direct Sampling method, **a** data-event, **b** training Image

systems, where the statistical parameters differ from one region to another [23, 96–98]. Some of the pixel-based, pattern-based, and hybrid MPS techniques are listed below:

Pixel-based MPS Methods

- *Extended Normal Equation Simulation (ENESIM)*: The ENESIM was the first method when the idea of MPS was introduced [99]. Extended indicator kriging is the main idea of the ENESIM which enables it to reproduce multiple-event derived from a training image. In this method, first the data event is located at each unknown point. Then, the training image is scanned to find all matchings. Using all matches, it constructs a conditional distribution for all occurrences. The value of the unknown point is sampled from the histogram generated using the conditional distribution. This is the main bottleneck of this method as it needs to scan the entire training image and the simulation grid to construct the conditional distribution for each visiting point. As a result, it is not practical to use this method on large G and TI.

- *Singe Normal Equation Simulation (SNESIM)*: The SNESIM was introduced to address the shortcomings of the original method, ENESIM, with the help of search trees. These search trees resolved the need for rescanning the entire TI each time visiting an unknown point. Instead, the searching task can be done once before the simulation, and the values of frequency or probability of the patterns is stored in the search tree [89] resulting in reduced computational cost. The probabilities in the search-tree are retrieved according to the pattern of the data in the data-event whenever an unknown point is visited. Since the SNESIM is a pixel-based method, it perfectly honors the hard/conditioning data. One of the advantages of the SNESIM is that it can use multiple-grid by which, at first, fewer nodes are used to capture the large structures after which the details are added. The major drawback of the SNESIM is that it is limited to modeling of categorical data such as facies modeling. Also, it is inefficient for modeling large models with millions of cells [24]. Later on, several methods were proposed upon the improvement of the efficiency and quality of the SNESIM [100, 101]. For more information on the SNESIM please refer to [65, 102].

- *Direct Sampling (DS)*: The direct sampling is introduced in recent years [23]. The advantage of the DS over ENESIM and SNESIM is that it can be used both for categorical and continuous variables with no limitations. In this method, only a fraction of the TI is scanned, and if any occurrence is found, it is directly copied to the visiting point in the simulation grid. No search-tree is constructed and the training image is scanned at each loop. Therefore, it requires less memory than the SNESIM. Like other MPS methods, the job of finding the similar patterns is done by the use of a distance function between the data-event and the training image. To elaborate more, in this algorithm, an unknown point is selected and the distance between the data-event and the training image is computed using the predefined distance function. Whenever a matching data-event in the training image is occurred (distance is less than the predefined threshold), the value of the central point of the data-event in the training image is pasted into the visiting point. If no match is found based on the predefined threshold, the most similar pattern found in the predefined portion of the training image is used. Figure 2.3 shows the schematic of the DS method. In this figure, the data-event in the search grid (Fig. 2.3a) is searched for in the training image and a match is found. Therefore, the color of the unknown point should be yellow. There are also extensions to this method. Bunch Direct Sampling and Advanced Direct Sampling are two of the extensions to the original direct sampling. In the advanced DS, the size of the data-event varies during the simulation [85]. In the Bunch DS, instead of a single node, a bunch of pixels are pasted to the matching region [103].

- *Cumulants*: This method, first introduced in [104], has the ability to directly extract the information beyond two-point statistics rather than the training image. This method first extracts the conditional probabilities from the available data. Searching the TI is only done if sufficient replicates are not found in the data. The limitation of this approach is that there is not a specific strategy to choose proper spatial cumulants for the geological scenarios.

Pattern-based MPS Methods

- Geological structures' continuity cannot be well preserved by pixel-based methods. Pattern-based methods are developed to address this issue. The difference between the pattern-based and pixel-based methods is that pattern-based methods paste an entire patch in the simulation grid instead of a single pixel. Some of the pattern-based MPS methods are explained in the following. A primary goal of implementing pattern-based methods is that they can preserve the geologic realism in the training image.
- *Simulation of Pattern (SIMPAT)*: To tackle the issues related to the CPU time and pattern connectivity in the SNESIM method, the SIMPAT algorithm was introduced by Arpat [105]. Probabilities are replaced by a distance measure to find occurrences of similar patterns. In this method, the patterns are extracted from the training image using a predefined template and stored in a database. Then, a random path is generated and the unknown points are visited following the generated path to extract the data-events. If there is no data in the data-event, a random pattern is selected from the database; otherwise, the pattern is selected based on the similarity of the data-event pattern to the patterns in the database. This method gives realistic realizations, but it suffers from high computational expenses, especially when dealing with three-dimensional models. This is because it needs to calculate the distance between the data-event and all patterns in the database. There are some extensions to the SIMPAT algorithm, such as flavoring it with a Bayesian framework [106] and incorporating wavelet decomposition [107].
- *Filter-based Simulation (FILTERSIM)*: The FILTERSIM was introduced by Zhang et al. [108] to resolve the issues related to the high computational cost of the SIMPAT. In this method, different filters are used with the purpose of reducing

the spatial complexity and dimensions. This method involves a preliminary step of filtering the patterns by linear filters. Then, the filtered patterns are clustered. The average of the patterns in each cluster is computed and used to calculate the distance function. The average pattern is also known as the prototype pattern. The most similar prototype pattern to the data-event based on the distance function is found and a random pattern from the pattern cluster corresponding to the prototype pattern is selected. The limited number of filters in the FILTERSIM makes it computationally cheaper than the SIMPAT. The limitations of the FILTERSIM are the need for selecting proper parameters for each training image and the use of linear filters that hinders it from extracting all the variability and information in the training image. Gloaguen and Dimitrakopoulos [109] replaced the distance function with wavelet. However, the wavelet-based methods have lots of parameters that need to be tuned to achieve good results, which can increase the CPU time and the quality of the results.

- *Cross-Correlation-based Simulation (CCSIM)*: The CCSIM was introduced by Tahmasebi et al. [110] in which the probability function was replaced by a cross-correlation function (CCF). This function is used to find the occurrences of pattern matches of data-event and the training image. An improvement on this method, named multi-scale CCSIM (MS-CCSIM) [24], was proposed to make it applicable on very large simulation grids. However, this method suffers from a poor match of well data and creation of some artifacts in the neighboring regions of the conditioning data.

Hybrid Methods

- As discussed above, each MPS method has its own pros and cons. Pixel-based algorithms were able to honor the hard data; however, they were unable to capture the large-scale connectivities. On the other hand, pattern-based methods were good at producing large-scale structures, but had difficulty in preserving the conditioning data. Therefore, some hybrid MPS techniques were introduced that take benefit from the advantages of both groups. *Hybrid Pixel- and Pattern-based Simulation (HYPPS)* [24] and *Hybrid Sequential Gaussian/Indicator Simulation and TI* [111] are two of the hybrid MPS methods. Explaining the details of these methods is beyond the scope of this book. Interested readers can refer to the cited references.

MPS methods have been widely used by different authors to generate reservoir model realizations. Azamifard et al. [112] used six MPS methods (Advanced Direct Sampling, Bunch Direct Sampling, Graph-cut, PCTOSIM, Image Quilting, MS-CCSIM) and one kriging-based method (SGSIM) to generate permeability realizations under real reservoir conditions and injection rates. The above geostatistical methods were assessed in terms of generating permeability fields enduring within the lifespan of the reservoir. The methods were assessed based on trapping and bypassing of oil, variogram, and histogram of permeability values, cumulative oil production, and oil production rate of 25 realizations generated by each method and the reference case. According to the results, most MPS methods showed superior performance to SGSIM in terms of better match of water front movement, wider uncertainty span (see Figs. 14 and 19 through 21 in [112]), similar bypassing and trapping of oil as in the reference case, and better oil production/injection matches. All methods preserved the variogram of the reference case and the training image indicating that the variogram is not a good comparative criterion. The SGSIM and PCTOSIM showed a narrower range of uncertainty and smoother permeability maps.

In another work, Azamifard et al. [113] employed two MPS methods, including the MS-CCSIM and Image Quilting to study the ability of MPS methods in generating permeability fields that can correctly predict oil production and water front with different oil viscosities in EOR projects. The ability of the MPS methods to generate both long and short variations of the correlation-length of permeability was their main reason of selecting this category of methods for heterogeneous reservoir characterization. Results showed that the MPS methods could mimic the flanging and fingering of water through the reservoir simulation even with different mobility ratios.

2.8.4 Challenges in MPS Methods

According to the cited applications of different MPS methods, these methods have shown acceptable performance at dealing with complex and realistic problems, in which, the geologic realism and hard data are honored. Nonetheless, the MPS techniques still encounter some critical challenges that keep the researches open in this field:

- **It is hard to condition the model to hard/secondary datasets** (e.g., seismic and well data): Mostly, Pattern-based algorithms suffer from this issue since these algorithms require computing the distance between the patterns in the training image and the model and between the data that is needed to be honored precisely in conditional simulation [114–116].
- **The low similarity between the internal patterns in the training image**: It is very challenging to prepare a comprehensive and distinctive training image. Thereby, the created structures by the MPS methods with small TIs are prone to discontinuities and unrealistic features. However, if the training image is very large, the computational cost of the methods increases significantly.
- **Similarity functions are not sufficient enough**: Distance functions currently being used in MPS methods cannot use higher order statistical moments and are based on two-point statistics, which cannot identify optimal patterns in the training image.
- **Lack of more realistic validation approaches**: There are a few number of validation methods that can be used to assess the reliability of newly developed MPS methods [117]. For instance, there can be significant difference between the realizations and the training image, while current methods cannot be used to quantify

it. As a result, visual comparison is the best method to verify the performance and reliability of the MPS techniques.

- **Stationarity of training images**: current MPS methods assume that the properties of interest are stationary. However, most of large-scale reservoirs have non-stationary properties. Wu et al. proposed two approaches, including the local rotation/affinity and the region concept, to resolve the non-stationarity constraints in SNESIM [118]. Local rotation/affinity is applicable to local non-stationarity whenever the non-stationarity in geological structures is related to orientations and sizes of structures. Therefore, it could only mitigate the non-stationary structures that could be made stationary by rotation and affinity. To do so, multiple rotation and affinity regions were defined and the transformations were done in each region to make it stationary. Finally, the simulation was done within each region using a search template and specific tree for that region. This approach is memory-demanding due to the high number of search trees stored in memory. The region concept was proposed to take into account the 'general' non-stationarity when the geological patterns vary from one sub-region to another sub-region within a domain. To tackle this kind of non-stationarity, different training images were used for each individual sub-region. This solution is CPU-demanding since the simulation must be repeated for each region with a different TI [118]. Some other attempts were made to resolve the non-stationarity issue that can be found in [97, 98, 119].

So far, we have discussed about the geological uncertainty and its types and the methods of generating geological realizations. However, their application in the process of reservoir management under uncertainty is not yet provided. Although these realizations give a preliminary representation of the underground reservoir, we are not certain about their accuracy. Therefore, they should be conditioned to production data; otherwise, a large number of realizations should be generated to cover the range of uncertainty in the model parameters.

In the next chapter, we will discuss about reducing the range of uncertainty by assimilating new observed (production) data from the reservoir, which is known as history matching. Different types of data used in history matching and their scales are discussed, challenges in history matching are introduced, different approaches to history matching are presented, and their pros and cons are provided.

References

1. Kayode O, Engineering G, Science A (2018) Multilateral well modeling from compartmentalized reservoirs
2. Anderson MP, Woessner WW, Hunt RJ (2015) Applied groundwater modelling: simulation of flow and advective transport, 2nd edn. Academic, New York
3. Troldborg L, Ondracek M, Koch J, Kidmose J, Refsgaard JC (2021) Quantifying stratigraphic uncertainty in groundwater modelling for infrastructure design. Hydrogeol J 29(3):1075–1089. https://doi.org/10.1007/s10040-021-02303-5

4. Refsgaard JC, van der Sluijs JP, Højberg AL, Vanrolleghem PA (2007) Uncertainty in the environmental modelling process—a framework and guidance. Environ Model Softw 22(11):1543–1556. https://doi.org/10.1016/j.envsoft.2007.02.004

5. Walker W, Harremoës P, Rotmans J, Van der Sluijs J, Van AM, Janssen P, von Krauss KKM (2003) Defining uncertainty a conceptual basis for uncertainty management in modelbased decision support. Integr Assess 4(1):5–17

6. Scheidt C, Zabalza-Mezghani I, Feraille M, Collombier D (2007) Toward a reliable quantification of uncertainty on production forecasts: adaptive experimental designs. Oil Gas Sci Tech 62(2 special issue):207–224. https://doi.org/10.2516/ogst:2007018

7. Shams M, El-Banbi A, Sayyouh H (2019) A novel assisted history matching workflow and its application in a full field reservoir simulation model. J Pet Sci Technol 9(3):64–87. https://doi.org/10.22078/jpst.2019.3407.1545

8. Lee K, Lim J, Ahn S, Kim J (2018) Feature extraction using a deep learning algorithm for uncertainty quantification of channelized reservoirs. J Petrol Sci Eng 171:1007–1022. https://doi.org/10.1016/j.petrol.2018.07.070

9. Kostakis FF, Mallison BT, Durlofsky LJ (2018) Multifidelity framework for uncertainty quantification with multiple quantities of interest. In: 16th European Conference on the Mathematics of Oil Recovery, ECMOR 2018. https://doi.org/10.3997/2214-4609.201802224

10. Ehlers L, Refsgaard J, Sonnenborg T (2019) Observational and predictive uncertainties for multiple variables in a spatially distributed hydrological model. Hydrol Process 33(33):833–848

11. Chakra NCC, Saraf DN (2016) History matching of petroleum reservoirs employing adaptive genetic algorithm. J Petrol Expl Product Tech 6(4):653–674. https://doi.org/10.1007/s13202-015-0216-4

12. Seifert D, Sonnenborg T, Refsgaard J, Højberg A, Troldborg L (2012) Assessment of hydrological model predictive ability given multiple conceptual geological models. Water Resour Res 48(6). https://doi.org/10.1029/2011WR011149

13. Selroos JO, Walker DD, Ström A, Gylling B, Follin S (2002) Comparison of alternative modelling approaches for groundwater flow in fractured rock. J Hydrol 257(1–4):174–188. https://doi.org/10.1016/S0022-1694(01)00551-0

14. Butts MB, Payne JT, Kristensen M, Madsen H (2004) An evaluation of the impact of model structure on hydrological modelling uncertainty for streamflow simulation. J Hydrol 298(1–4):242–266. https://doi.org/10.1016/j.jhydrol.2004.03.042

15. Silva MI de O, dos Santos AA de S, Schiozer DJ, de Neufville R (2017) Methodology to estimate the value of flexibility under endogenous and exogenous uncertainties. J Petrol Sci Eng 151:235–247. https://doi.org/10.1016/j.petrol.2016.12.026

16. Jahandideh A, Jafarpour B (2018) Stochastic oilfield optimization for hedging against uncertain future development plans. In: 16th European Conference on the Mathematics of Oil Recovery, ECMOR, 22–26. https://doi.org/10.3997/2214-4609.201802225

17. Bukshtynov V, Volkov O, Durlofsky LJ, Aziz K (2015) Comprehensive framework for gradient-based optimization in closed-loop reservoir management. Comput Geosci 19(4):877–897. https://doi.org/10.1007/s10596-015-9496-5

18. Jahandideh A, Jafarpour B (2018) Hedging against uncertain future development plans in closed-loop field development optimization. In: Proceedings—SPE Annual Technical Conference and Exhibition. https://doi.org/10.2118/191622-ms

19. He X, Koch J, Sonnenborg TO, Flemming J, Schamper C, Refsgaard JC (2014) Using airborne geophysical data and borehole data. Water Resour Res,1–23. https://doi.org/10.1002/2013WR014593.Received

20. Koch J, He X, Jensen KH, Refsgaard JC (2014) Challenges in conditioning a stochastic geological model of a heterogeneous glacial aquifer to a comprehensive soft data set. Hydrol Earth Syst Sci 18(8):2907–2923. https://doi.org/10.5194/hess-18-2907-2014

21. Strebelle S (2002) Conditional simulation of complex geological structures using multiple-point statistics. Math Geol 34(1):1–21. https://doi.org/10.1023/A:1014009426274

22. Mariethoz G, Caers J (2014) Multiple-point geostatistics: stochastic modeling with training images. Wiley Blackwell. https://doi.org/10.1002/9781118662953
23. Mariethoz G, Renard P, Straubhaar J (2010) The direct sampling method to perform multiple-point geostatistical simulations. Water Resour Res 46(11). https://doi.org/10.1029/2008WR007621
24. Tahmasebi P, Sahimi M, Caers J (2014) MS-CCSIM: accelerating pattern-based geostatistical simulation of categorical variables using a multi-scale search in Fourier space. Comput Geosci 67:75–88. https://doi.org/10.1016/J.CAGEO.2014.03.009
25. Aydin O, Caers JK (2017) Quantifying structural uncertainty on fault networks using a marked point process within a Bayesian framework. Tectonophysics 712–713:101–124. https://doi.org/10.1016/j.tecto.2017.04.027
26. Li Z, Wang X, Wang H, Liang RY (2016) Quantifying stratigraphic uncertainties by stochastic simulation techniques based on Markov random field. Eng Geol 201:106–122. https://doi.org/10.1016/j.enggeo.2015.12.017
27. Maschio C, Schiozer DJ (2014) Bayesian history matching using artificial neural network and Markov Chain Monte Carlo. J Petrol Sci Eng 123:62–71. https://doi.org/10.1016/J.PETROL.2014.05.016
28. Thore P, Shtuka A, Lecour M, Ait-Ettajer T, Cognot R (2002) Structural uncertainties: determination, management, and applications. Geophysics 67(3):840–852. https://doi.org/10.1190/1.1484528
29. Sundell J, Rosén L, Norberg T, Haaf E (2016) A probabilistic approach to soil layer and bedrock-level modeling for risk assessment of groundwater drawdown induced land subsidence. Eng Geol 203:126–139. https://doi.org/10.1016/j.enggeo.2015.11.006
30. Xiao T, Zhang L-M, Li X-Y, Li D-Q (2017) Probabilistic stratification modeling in geotechnical site characterization. ASCE-ASME J Risk Uncertainty Eng Syst, Part A: Civil Engineering. 3(4):1–10. https://doi.org/10.1061/ajrua6.0000924
31. Gong W, Tang H, Wang H, Wang X, Juang CH (2018) Probabilistic analysis and design of stabilizing piles in slope considering stratigraphic uncertainty. Eng Geol 2019(259):105162. https://doi.org/10.1016/j.enggeo.2019.105162
32. Mukerji T, Jørstad A, Avseth P, Mavko G, Granli JR (2012) Mapping lithofacies and pore-fluid probabilities in a North Sea reservoir: seismic inversions and statistical rock physics. Geophysics 66(4):988–1001. https://doi.org/10.1190/1.1487078
33. Rosenbaum MS, Culshaw MG (2003) Communicating the risks arising from geohazards. J R Stat Soc A Stat Soc 166(2):261–270. https://doi.org/10.1111/1467-985X.00275
34. Budd AF, Petersen RA, McNeill DF (1998) Stepwise Faunal change during evolutionary turnover: a case study from the Neogene of Curaçao, Netherlands Antilles. PALAIOS. 13(2):170–188. https://doi.org/10.2307/3515488
35. Potts DC (1984) Generation times and the quaternary evolution of reef-building corals. Paleobiol 10(1):48–58. http://www.jstor.org/stable/2400500
36. Pjasek RB (1980) Toxic and hazardous waste disposal. Butterworth-Heinemann
37. Wood R, Curtis A (2004) Geological prior information and its applications to geoscientific problems. Geol Soc Spec Pub 239:1–14. https://doi.org/10.1144/GSL.SP.2004.239.01.01
38. Lochbühler T, Pirot G, Straubhaar J, Linde N (2014) Conditioning of multiple-point statistics facies simulations to tomographic images. Math Geosci 46(5):625–645. https://doi.org/10.1007/s11004-013-9484-z
39. Hernandez-Martinez E, Perez-Muñoz T, Velasco-Hernandez JX, Altamira-Areyan A, Velasquillo-Martinez L (2013) Facies recognition using multifractal Hurst analysis: applications to well-log data. Math Geosci 45(4):471–486. https://doi.org/10.1007/s11004-013-9445-6
40. Fang J, Gong B, Caers J (2022) Data-driven model falsification and uncertainty quantification for fractured reservoirs. Engineering. https://doi.org/10.1016/j.eng.2022.04.015
41. Adekunle SO, Desmond O, Nwakanma A (2022) Quantitative interpretation of petrophysical parameters for reservoir characterization in an Onshore Field of Niger Delta Basin. J Geograp, Environ Earth Sci Inter 33(April):14–36. https://doi.org/10.9734/jgeesi/2022/v26i330339

42. Nawaz MA, Curtis A (2017) Bayesian inversion of seismic attributes for geological facies using a Hidden Markov Model. Geophys J Int 208(2):1184–1200. https://doi.org/10.1093/gji/ggw411

43. Grana D (2013) Bayesian inversion methods for seismic reservoir characterization and time-lapse studies. Stanford University

44. Azevedo L, Nunes R, Correia P, Soares A, Neto GS, Guerreiro L (2019) Stochastic direct facies seismic AVO inversion. In: Society of Exploration Geophysicists International Exposition and 83rd Annual Meeting, SEG 2013: Expanding Geophysical Frontiers, 2352–2356. https://doi.org/10.1190/segam2013-0555.1

45. Connolly PA, Hughes MJ (2016) Stochastic inversion by matching to large numbers of pseudo-wells. Geophysics 81(2):M7–M22. https://doi.org/10.1190/GEO2015-0348.1

46. Beucher H, Renard D (2016) Truncated Gaussian and derived methods. CR Geosci 348(7):510–519. https://doi.org/10.1016/J.CRTE.2015.10.004

47. Bougher BB, Herrmann FJ (2016) AVA classification as an unsupervised machine-learning problem. SEG Tech Program Expand Abs 35:553–556. https://doi.org/10.1190/segam2016-13874419.1

48. Scales JA, Tenorio L (2001) Prior information and uncertainty in inverse problems. Geophysics 66(2):389–397. https://doi.org/10.1190/1.1444930

49. Buland A, Henning O (2020) Bayesian AVO inversion. Geophysics 68(1):185–198. https://doi.org/10.3997/2214-4609-pdb.28.p155

50. Grana D, Fjeldstad T, Omre H (2017) Bayesian Gaussian mixture linear inversion for geophysical inverse problems. Math Geosci 49(4):493–515. https://doi.org/10.1007/s11004-016-9671-9

51. Costa E, Silva Talarico E, Grana D, Passos De Figueiredo L, Pesco S (2020) Uncertainty quantification in seismic facies inversion. Geophysics 85(4):M43–M56. https://doi.org/10.1190/geo2019-0392.1

52. Doyen P (2007) Seismic reservoir characterization: an earth modelling perspective. EAGE Publications

53. Cremon MA, Christie M, Gerritsen MG (2019) Monte Carlo simulation for uncertainty quantification in reservoir simulation: a convergence study. In: 16th European Conference on the Mathematics of Oil Recovery, ECMOR 2018, 190, December. https://doi.org/10.3997/2214-4609.201802226

54. Hastings WK (1970) Monte carlo sampling methods using Markov chains and their applications. Biometrika 57(1):97–109. https://doi.org/10.1093/biomet/57.1.97

55. Goodfellow I, Bengio Y, Courville A (2018) Deep learning. The MIT Press

56. Metropolis N, Rosenbluth AW, Rosenbluth MN, Teller AH, Teller E (1953) Equation of state calculations by fast computing machines. J Chem Phys 21(6):1087–1092. https://doi.org/10.1063/1.1699114

57. Lu Z, Zhang D (2003) On importance sampling Monte Carlo approach to uncertainty analysis for flow and transport in porous media. Adv Water Resour 26(11):1177–1188. https://doi.org/10.1016/S0309-1708(03)00106-4

58. Abdollahzadeh A, Christie M, Corne D (2015) An adaptive metropolis-hasting sampling algorithm for reservoir uncertainty quantification in Bayesian inference. Society of Petrol Eng—SPE Reserv Simul Symposium 2015(2):1243–1265. https://doi.org/10.2118/173263-MS

59. Fan YR, Shi X, Duan QY, Yu L (2022) Towards reliable uncertainty quantification for hydrologic predictions, Part I: development of a particle copula Metropolis Hastings method. J Hydrol 612, July. https://doi.org/10.1016/j.jhydrol.2022.128163

60. Zhao M, Huang Y, Zhou W, Li H (2021) Bayesian uncertainty quantification for guided-wave-based multidamage localization in plate-like structures using Gibbs sampling. Struct Health Monit 20(6):3092–3112. https://doi.org/10.1177/1475921720979352

61. Langmore L, Dikovsky M, Geraedts S, Norgaard P, von Behren R (2023) Hamiltonian Monte Carlo in inverse problems. Ill-conditioned and multimodality. Int J Uncertain Quant 13(1):69–93. https://doi.org/10.1615/Int.J.UncertaintyQuantification.2022038478

62. Fischer A, Igel C (2011) Bounding the bias of contrastive divergence learning. Neural Comput 23(3):664–673. https://doi.org/10.1162/NECO_a_00085
63. Karakida R, Okada M, Amari S ichi. Dynamical analysis of contrastive divergence learning: restricted Boltzmann machines with Gaussian visible units. Neural Netw 79:78–87. https://doi.org/10.1016/j.neunet.2016.03.013
64. Tieleman T, Hinton G (2009) Using fast weights to improve persistent contrastive divergence. ACM Inter Confer Proceed Series. 382(8):1033–1040. https://doi.org/10.1145/1553374.155 3506
65. Liu Y, Sun W, Durlofsky LJ (2019) A deep-learning-based geological parameterization for history matching complex models. Math Geosci 51(6):725–766. https://doi.org/10.1007/s11 004-019-09794-9
66. Vo HX, Durlofsky LJ (2015) Data assimilation and uncertainty assessment for complex geological models using a new PCA-based parameterization. Comput Geosci 19(4):747–767. https://doi.org/10.1007/s10596-015-9483-x
67. Carlo M. Water Resources Research (1969) J Am Water Resour Assoc 5(3):2–2. https://doi.org/10.1111/j.1752-1688.1969.tb04897.x
68. Chen B, He J, Wen XH, Chen W, Reynolds AC (2017) Uncertainty quantification and value of information assessment using proxies and Markov chain Monte Carlo method for a pilot project. J Petrol Sci Eng 157:328–339. https://doi.org/10.1016/j.petrol.2017.07.039
69. Hu LY (2000) Gradual deformation and iterative calibration of Gaussian-related stochastic models. Math Geol 32(1):87–108. https://doi.org/10.1023/A:1007506918588
70. Shi G (2014) Kriging. In: Data mining and knowledge discovery for geoscientists. Elsevier, pp 238–274. https://doi.org/10.1016/B978-0-12-410437-2.00008-4
71. Remy N, Boucher A, Wu J (2011) Applied geostatistics with SGeMS: a user's guide
72. Deutsch CV (2003) Geostatistics. Academic Press. https://doi.org/10.1016/B0-12-227410-5/00869-3
73. Intrinsic MG, Functions R, Applications T (1973) Adv Appl Probab 5(3):439–468. https://doi.org/10.1017/S0001867800039379
74. Lichtenstern A (2013) Kriging methods in spatial statistics. TUM Media Online, p 97
75. Journel A, Zhang T (2006) The necessity of a multiple-point prior model. undefined 38(5):591–610. https://doi.org/10.1007/S11004-006-9031-2
76. Journel AG, Huijbregts CJ (2003) Mining geostatistics. Blackburn Press
77. Awad A, Bazan P, German R (2014) SGsim: a simulation framework for smart grid applications. In: ENERGYCON 2014—IEEE International Energy Conference, pp 730–736. https://doi.org/10.1109/ENERGYCON.2014.6850507
78. Matheron G, Beucher H, de Fouquet C, Galli A, Guerillot D, Ravenne C (1987) Conditional simulation of the geometry of Fluvio-Deltaic reservoirs. SPE J 27:123–131. https://doi.org/10.2118/16753-MS
79. Albert F (1987) Stochastic imaging of spatial distributions using hard and soft information. Stanford University
80. Daya SBS, Qiuming C, Agterberg F (eds) (2018) Multiple point statistics: a review. In: Handbook of Mathematical Geosciences Fifty Years of IAMG. Springer Open, pp 613–643
81. Bai T, Tahmasebi P (2022) Sequential Gaussian simulation for geosystems modeling: a machine learning approach. Geosci Front 13(1):101258. https://doi.org/10.1016/J.GSF.2021.101258
82. Holden L, Hauge R, Skare Ø, Skorstad A (1998) Modeling of fluvial reservoirs with object models. Math Geol 30(5):473–496. https://doi.org/10.1023/A:1021769526425
83. Haldorsen HH, Damsleth E (1990) Stochastic modeling (includes associated papers 21255 and 21299). J Petrol Technol 42(04):404–412. https://doi.org/10.2118/20321-PA
84. Renard P, Mariethoz G, Comunian A, Straubhaar J (2014) Hybrid geostatistics: Object-based simulations using MPS-generated meandering channels. In: 2nd EAGE Integrated Reservoir Modelling Conference—Uncertainty Management: Are we Doing it Right?. https://doi.org/10.3997/2214-4609.20147474

85. Meerschman E, Pirot G, Mariethoz G, Straubhaar J, Van Meirvenne M, Renard P (2013) A practical guide to performing multiple-point statistical simulations with the direct sampling algorithm. Comput Geosci 52:307–324. https://doi.org/10.1016/j.cageo.2012.09.019
86. Gómez-Hernández JJ, Wen XH (1998) To be or not to be multi-Gaussian? a reflection on stochastic hydrogeology. Adv Water Resour 21(1):47–61. https://doi.org/10.1016/S0309-170 8(96)00031-0
87. Bianchi M, Zheng C, Wilson C, Tick GR, Liu G, Gorelick SM (2011) Spatial connectivity in a highly heterogeneous aquifer: from cores to preferential flow paths. Undefined 47(5). https://doi.org/10.1029/2009WR008966
88. Renard P, Allard D (2013) Connectivity metrics for subsurface flow and transport. Adv Water Resour 51:168–196. https://doi.org/10.1016/J.ADVWATRES.2011.12.001
89. Sagar BSD, Cheng Q, Agterberg F (2018). Handbook of Math Geosci. https://doi.org/10. 1007/978-3-319-78999-6
90. Pickel A, Frechette JD, Comunian A, Weissmann GS (2015) Building a better training image with digital outcrop models. J Hydrol. https://doi.org/10.1016/j.jhydrol.2015.08.049
91. Lantuéjoul C (2002) Introduction. In: Geostatistical simulation. Springer, Berlin, Heidelberg, pp 1–6. https://doi.org/10.1007/978-3-662-04808-5_1
92. Deutsch CV, Tran TT (2002) FLUVSIM: a program for object-based stochastic modeling of fluvial depositional systems. Comput Geosci 28(4):525–535. https://doi.org/10.1016/S0098-3004(01)00075-9
93. Seminara G (2006) Meanders. J Fluid Mech 554:271–297. https://doi.org/10.1017/S00221 12006008925
94. Pyrcz MJ, Boisvert JB, Deutsch CV (2009) ALLUVSIM: a program for event-based stochastic modeling of fluvial depositional systems. Comput Geosci 35(8):1671–1685. https://doi.org/ 10.1016/J.CAGEO.2008.09.012
95. Biswal B, Øren PE, Held RJ, Bakke S, Hilfer R (2007) Stochastic multiscale model for carbonate rocks. Phys Rev E Stat Nonlinear Soft Matter Phys 75(6):061303. https://doi.org/ 10.1103/PHYSREVE.75.061303/FIGURES/4/MEDIUM
96. Tahmasebi P, Sahimi M (2015) Reconstruction of nonstationary disordered materials and media: Watershed transform and cross-correlation function. Phys Rev E—Statis, Nonlinear, Soft Matter Phys 91(3). https://doi.org/10.1103/PHYSREVE.91.032401/FIG URES/16/MEDIUM
97. Honarkhah M, Caers J (2012) Direct pattern-based simulation of non-stationary geostatistical models. Math Geosci 44(6):651–672. https://doi.org/10.1007/S11004-012-9413-6
98. Chugunova TL, Hu LY (2008) Multiple-point simulations constrained by continuous auxiliary data. Math Geosci 40(2):133–146. https://doi.org/10.1007/S11004-007-9142-4
99. Guardiano FB, Srivastava RM (1993) Multivariate geostatistics: beyond bivariate moments. In: Geostatistics Troia '92, Vol. 1. Springer, Dordrecht, pp 133–144. https://doi.org/10.1007/ 978-94-011-1739-5_12
100. Straubhaar J, Walgenwitz A, Renard P (2013) Parallel multiple-point statistics algorithm based on list and tree structures. Math Geosci 45:131–147. https://doi.org/10.1007/s11004-012-9437-y
101. Cordua KS, Hansen TM, Mosegaard K (2014) Improving the pattern reproducibility of multiple-point-based prior models using frequency matching. Math Geosci. https://doi.org/ 10.1007/s11004-014-9531-4
102. Boucher A (2009) Considering complex training images with search tree partitioning. Comput Geosci 35(6):1151–1158. https://doi.org/10.1016/J.CAGEO.2008.03.011
103. Rezaee H, Mariethoz G, Koneshloo M, Asghari O (2013) Multiple-point geostatistical simulation using the bunch-pasting direct sampling method. Comput Geosci 54:293–308. https:// doi.org/10.1016/J.CAGEO.2013.01.020
104. Dimitrakopoulos R, Mustapha H, Gloaguen E (2009) High-order statistics of spatial random fields: exploring spatial cumulants for modeling complex non-Gaussian and non-linear phenomena. Math Geosci 42(1):65–99. https://doi.org/10.1007/S11004-009-9258-9
105. Arpat B (2005) Sequential simulation with patterns. Stanford University

106. Abdollahifard MJ, Faez K (2013) Stochastic simulation of patterns using Bayesian pattern modeling. Comput Geosci 17(1):99–116. https://doi.org/10.1007/s10596-012-9319-x

107. Mustapha H, Dimitrakopoulos R (2010) High-order stochastic simulation of complex Spatially distributed natural phenomena. Math Geosci 42(5):457–485. https://doi.org/10.1007/s11004-010-9291-8

108. Zhang T, Switzer P, Journel A (2006) Filter-based classification of training image patterns for spatial simulation. Math Geol 38(1):63–80. https://doi.org/10.1007/S11004-005-9004-X

109. Gloaguen E, Dimitrakopoulos R (2009) Two-dimensional conditional simulations based on the wavelet decomposition of training images. Math Geosci 41(6 special issue):679–701. https://doi.org/10.1007/s11004-009-9235-3

110. Tahmasebi P, Hezarkhani A, Sahimi M (2012) Multiple-point geostatistical modeling based on the cross-correlation functions. Compu Geosci 16(3):779–797. https://doi.org/10.1007/S10596-012-9287-1

111. Ortiz JM, Deutsch CV (2004) Indicator simulation accounting for multiple-point statistics. Math Geol 36(5):545–565. https://doi.org/10.1023/B:MATG.0000037736.00489.B5

112. Azamifard A, Rashidi F, Pourfard M, Ahmadi M, Dabir B (2020) Enduring effect of permeability texture for enhancing accuracy and reducing uncertainty of reservoir fluid flow through porous media. Pet Sci 17(1):118–135. https://doi.org/10.1007/s12182-019-00366-4

113. Azamifard A, Ahmadi M, Rashidi F, Pourfard M, Dabir B (2020) Insights of new-generation reservoir property modeling (MPS methods) in assessing the reservoir performance for different recovery methods. Arabian J Geosci 13(7). https://doi.org/10.1007/s12517-020-05293-y

114. Hashemi S, Javaherian A, Ataee-pour M, Tahmasebi P, Khoshdel H (2014) Channel characterization using multiple-point geostatistics, neural network, and modern analogy: a case study from a carbonate reservoir, southwest Iran. J Appl Geophys 111:47–58. https://doi.org/10.1016/j.jappgeo.2014.09.015

115. Rezaee H, Marcotte D (2017) Integration of multiple soft data sets in MPS thru multinomial logistic regression: a case study of gas hydrates. Stoch Env Res Risk Assess 31(7):1727–1745. https://doi.org/10.1007/s00477-016-1277-8

116. Tahmasebi P, Sahimi M (2015) Geostatistical simulation and reconstruction of porous media by a cross-correlation function and integration of hard and soft data. Trans Porous Media 107(3):871–905. https://doi.org/10.1007/s11242-015-0471-3

117. Tan X, Tahmasebi P, Caers J (2014) Comparing training-image based algorithms using an analysis of distance. Math Geosci 46. https://doi.org/10.1007/s11004-013-9482-1

118. Wu J, Zhang T, Boucher A (2007) Non-stationary multiple-point geostatistical simulations with region concept. In: Proceedings of the 20th SCRF meeting, Stanford, CA, USA, May, pp 1–53

119. Tahmasebi P, Sahimi M (2015) Reconstruction of nonstationary disordered materials and media: watershed transform and cross-correlation function. Phys Rev E 91:32421. https://doi.org/10.1103/PhysRevE.91.032401

Chapter 3
Reducing the Geological Uncertainty by History Matching

Abstract Whenever there are observed dynamic data obtained from the reservoir understudy, we can reduce the geological uncertainty by conditioning the prior geological realizations to the observed data (Oliver and Chen in Computational Geosciences. 15:185–221, 2010; Ghoniem et al. in Applied Mathematical Modelling. 8:282–287, 1984; Heidari et al. in Computers and Geosciences. 55:84–95, 2013;Zhang and Oliver in SPE Journal. 16:307–317, 2011). This kind of uncertainty management is an inverse uncertainty management/quantification, which is mainly based on Bayesian approaches as described in Sect. 1.5.2. Inverse uncertainty management is technically known as model updating, model conditioning, model adjusting, model calibration, parameter estimation, data assimilation, history matching, automated history matching, or computer-assisted history matching. No matter what you call this process, it helps to have better and more reliable estimations of the true model. This chapter provides the details of history matching, different data types and their scale that are used in history matching, use of seismic, static, and production data in history matching, challenges encountered during history matching, history matching methods, and different approaches to reservoir management under geological uncertainty. Open-loop and closed-loop reservoir management are described and their pros and cons are discussed. Also, the most commonly used ensemble-based methods, ensemble-smoother methods, and stochastic optimization algorithms used for history matching are described.

Keywords History matching · Seismic data · Static data · Dynamic data · Closed-loop reservoir management · Open-loop reservoir management · Ensemble-smoother · Stochastic optimization

3.1 Introduction to History Matching

History matching is the act of modifying reservoir properties in such a way that the simulated production data match the observed production data under a similar production strategy. In this manner, the range of uncertainty in the output variables will be decreased as better models are obtained. For example, Fig. 3.1 demonstrates the water-cut predictions in a well using the prior (unconditioned) and history

© The Author(s), under exclusive license to Springer Nature Switzerland AG 2023   43
R. Yousefzadeh et al., *Introduction to Geological Uncertainty Management in Reservoir Characterization and Optimization*, SpringerBriefs in Petroleum Geoscience & Engineering, https://doi.org/10.1007/978-3-031-28079-5_3

Fig. 3.1 Prediction of field
water-cut based on prior
realizations (green lines),
history matched realizations
(blue lines), and the true data
(black line)

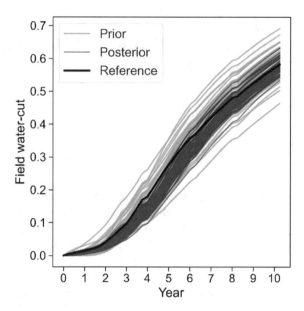

matched (conditioned) realizations of a reservoir model. In this figure, blue lines represent the water-cut obtained from the prior realizations, red lines represent the water-cut obtained from the history matched realizations, and the black dots show the actual observed data (reference). As can be seen from Fig. 3.1, the range of uncertainty is reduced after history matching.

History matching is an inverse problem [1], in which, the inputs and outputs are known, but the system that maps the inputs to outputs is unknown. Our goal is to find the underlying system, constrained to the geologic realism of the true model. However, the true model is not known, and we only have limited direct measurements of the reservoir properties along with observed production data, which are the response of the underground reservoir. In a petroleum reservoir, inputs include the well specifications, initial and boundary conditions, rock and fluid properties (e.g., porosity, permeability, fluid contacts, etc.), reservoir engineering properties (e.g., relative permeability, capillary pressure), and the outputs are the production data (pressure, oil rate, gas rate, etc.). However, we have to deal with a big degree of uncertainty in rock, fluid, and reservoir engineering properties. As a result, any prior model constructed based on the prior information from the underground reservoir is not certain. Since the number of unknown variables is higher than the number of known variables, history matching is an ill-conditioned inverse problem, in which, an infinite number of solutions can honor the historical production data. Consequently, it is not logical to rely on only one history-matched model to forecast the production in the future [5–7]. The process and methods of creating the prior geological realizations are described in Chap. 2.

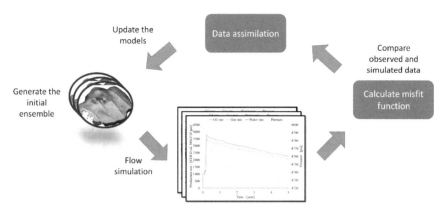

Fig. 3.2 History matching workflow

The schematic of history matching is illustrated in Fig. 3.2. In this procedure, first some prior reservoir models are built based on well correlations; seismic interpretations; geology, geophysics, and facies data; property modelling; and petrophysics. Then, the models are simulated using a reservoir simulator to obtain the simulated production values. If a match is obtained between the simulated and observed production data, the models can be used for predicting the future behavior of the reservoir. Otherwise, uncertain properties of the reservoir are modified until a match is achieved. According to above explanations, history matching is an optimization problem in which the decision variables are the uncertain reservoir properties and the objective function is the discrepancy between the simulated and observed production data. Equation 3.1 gives a general formulation of the objective function in history matching:

$$
J = \sum_{i=1}^{N_p} \sum_{j=1}^{N_t} w_i \left(\frac{x_i^{obs}\left(t^j\right) - x_i^{sim}\left(t^j\right)}{STD_i} \right)^2 , \tag{3.1}
$$

where N_p is the number of matching parameters (cumulative oil production, watercut, etc.), N_t is the total number of production samples (timesteps), w_i is the weight of the ith matching parameter, $x_i^{obs}\left(t^j\right)$ is the observed value of the ith parameter at the jth timestep, $x_i^{sim}\left(t^j\right)$ is the simulated value of the ith parameter at the jth timestep, and STD_i is the standard deviation of the ith matching parameter. That what matching parameters should be used for history matching can be found by sensitivity analysis [7].

The drawback of Eq. 3.1 is that it only measures the error between the simulated and actual production data. Care should be taken into account to not update the uncertain parameters in a direction that does not follow the geologic realism of the prior realizations. This constraint must be added to the updating rule of the data assimilation step. Therefore, the objective function can be re-formulated as:

$$J(m) = \frac{1}{2}\left(m - m_{prior}\right)^T C_M^{-1}\left(m - m_{prior}\right)$$
$$+ \frac{1}{2}(d_{obs} - g(m))^T C_D^{-1}(d_{obs} - g(m)),   \tag{3.2}$$

where m_{prior} is the prior model vector, m is the updated model vector, d_{obs} is the vector of observed output data, $g(m)$ is the simulation output on realization m, C_M^{-1} is the inverse of the covariance matrix of model parameters, and C_D^{-1} is the inverse of the covariance matrix of measurement errors in observed data. C_D is a diagonal matrix that its main diagonal represents the error variance. Therefore, measurement errors are regarded conditionally independent. The first group in Eq. 3.2 measures the deviation of the updated model from the prior realization, and the second group is the difference between the measured and simulated production data. It is usual to assign a weighting factor to the first group indicating its relative importance in calculating the objective function. If an ensemble of realizations has to be history matched, Eq. 3.2 is updated to:

$$J(m_i) = \frac{1}{2}\left(m_i - m_{i,prior}\right)^T C_M^{-1}\left(m_i - m_{i,prior}\right)$$
$$+ \frac{1}{2}(d_{obs} - g(m_i))^T C_D^{-1}(d_{obs} - g(m_i))   \tag{3.3}$$

for the ith member of the ensemble.

There are three main approaches that can be used for history matching:

1. Selecting an ensemble of prior realizations but conditioning only one of them to the observed production data (Fig. 3.3b). The model selected for updating can be the best matching model (Halliburton approach in [8]), or a random model from the ensemble (see [9] and Stanford/Chevron approach in [8]).
2. History matching of an ensemble of models (Fig. 3.3c). This approach is one of the most used approaches in history matching [10–18]. There are also different approaches to selecting the models for the production optimization stage, such as using the mean of the conditioned models [11, 13, 14], central model of the ensemble [16], model with maximum likelihood estimate [19], best updated model (see the Roxar method in [8]), or one of the matched models [20].
3. Uncertainty reduction using iterative approaches (Fig. 3.3d). This approach entails filtering the best matching prior models with the production data and discarding the others [21]. Accordingly, no data assimilation is done in this approach. Some authors have suggested to proceed with the best matched model to next steps (Schlumberger approach in [8], while others have suggested to use a set of representative realizations [22].

As stated in Chap. 1, information reliability is a source of uncertainty in reservoir management. Therefore, it is suggested to add a realistic noise to the observed data in order to account for this kind of uncertainty during data assimilation. This is also helpful to prevent the updated models from overfitting the measured data. Overfit

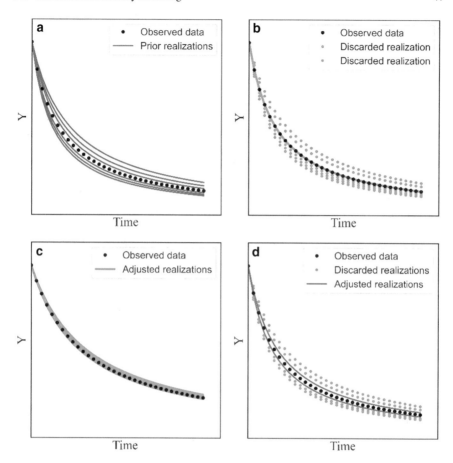

Fig. 3.3 Demonstration of the three main data assimilation approaches. **a** prior realizations, **b** using the best matching realization to the historical production data to adjust its parameters, **c** adjusting the parameters of all prior realizations, **d** accepting the best fitting prior models and discarding the others. In this figure, Y indicates an arbitrary dynamic performance indicator

models are able to perfectly reproduce the historical production data, but fail to provide reliable forecasts of the future. Therefore, they are not suitable to be used in the production optimization step.

3.2 Different Data Types and their Scale in History Matching

Generally, there are two groups of data used in history matching: 1- Static data, 2- Dynamic data.

Static data include the petrophysical data, such as facies, porosity, permeability, net-to-gross ratio; structural data, such as reservoir boundaries, layers top and bottom, fault, fracture network; stratigraphic data, such as stacking of layers, bed thickness and spacing, depositional process, fluid contacts, sediment source, fluid content; reservoir engineering data, such as fluid properties, aquifer strength, relative permeability, capillary pressure, hydraulic communication between layers, faults transmissibility multiplier, fluid contact levels, reservoir pressure, and fluid saturations; 2D and 3D seismic data, such as shear wave (S-wave) velocity, compression wave (P-wave) velocity and densities. These data are used to build the prior static models using the uncertainty quantification methods described in Chap. 2 [23]. Seismic data can be used to make relationship between well logs and core data on one hand, and well-test analysis and tracer on the other [24].

Dynamic data are those data obtained from the sub-surface reservoir or reservoir simulation as a function of time. These data are the response of the generated or the true model, which uncertain parameters are adjusted in such a way to get a match between the simulated and true value of these parameters. Dynamic data include the production data, well testing data, and 4D seismic data. Note that, 4D seismic data are 3D seismic surveys of reservoir and fluid properties over time, which is also known as time-lapse 3D seismic data.

Table 3.1 shows the scale of different data types introduced above.

3.3 Using Seismic, Static, and Production Data in History Matching

Seismic data are used in two ways in history matching: (1) building the static model, (2) data assimilation/history matching. Regarding that there are three types of seismic data, including two-, three-, and four-dimensional seismic data, 2D and 3D seismic data are used to build the static reservoir model, and 4D seismic data are used for data assimilation and history matching.

Since we cannot rely on only a single type of data to characterize the reservoir model, it is better to integrate data of various types, including 2D, 3D, 4D seismic data, petrophysical, well logs, core, and production data. Generally, the role of seismic data in building geological models and distributing reservoir properties away from the wells cannot be overlooked. As stated above, seismic data are beneficial in different reservoir characterization stages. For instance, seismic horizons can be interpreted to create structural maps of different reservoir formations [25, 26]. Data from seismic, well logs, and core can be combined to constructed facies models.

Table 3.1 Data types and their scale in history matching

Data type			Scale
Static data	Petrophysical data	Facies	< 20 (facies indicator)
		Porosity	0–1 (fraction)
		Permeability	0–10,000 mD
		Net-to-gross ratio	0–1 (fraction)
	Structural data	Reservoir boundaries	0–3 (boundary type)
		Layers top and bottom	100–10,000 feet
		Fault	0–1 (exist/not-exist)
		Fracture network	0–5 (network type)
	Stratigraphic data	Bed thickness and spacing	1–300 feet
		Depositional process	< 10 (deposition type)
		Fluid contacts	100–10,000 feet
		Sediment source	< 10 (source indicator)
		Fluid content	0–2 (oil, water, gas)
	Reservoir engineering data	Fluid properties	Compressibility: 1E-6–1E-3 Density: 0.001–100 lb/ft^3 Bubble/dew point: 1E+2–3E+3 Composition: 0–1 (fraction) Viscosity: 1E-1–1E+5 cp
		Aquifer strength	1–1000 (unitless)
		Relative permeability	0–1 (fraction)
		Capillary pressure	−100–10,000 psi
		Fault transmissibility multiplier	0–1 (fraction)
		Reservoir pressure	2000–8000 psi
		Fluid saturations	0–1 (fraction)
Dynamic data	Production data	Production/Injection rates	0–10,000 STB/D
		Cumulative production/injection	0–1E10 STB
		Water-cut	0–1 (fraction)
		Gas-oil/water-oil/water-gas ratio	0–2 MCSF/STB,
		Reservoir/bottom-hole pressure	0–8000 psi
	4D seismic data	P-wave velocity	340–5000 m/s
		S-wave velocity	200–3000 m/s

(continued)

Table 3.1 (continued)

Data type			Scale
		Rayleigh waves velocity	180–2700 m/s
		Density	30–200 lb/ft^3
		Bulk modulus	1E4–1E5 psi
Seismic data	2D, 3D, and 4D seismic data	As 4D seismic data	

Seismic attributes, like inverted acoustic impedance, are used as secondary data to constrain geostatistically generated reservoir properties (e.g., porosity, permeability, etc.) [27].

3.3.1 Two-dimensional Seismic Data

If the shot and receivers are aligned in a straight line, the obtained seismic data will be two-dimensional (Fig. 3.4). The next line of shot and receivers are placed kilometers away. This means that 2D seismic data are acquired independently contrary to 3D seismic data that are acquired by multiple closely aligned lines. 2D seismic is less expensive and has less environmental impact. 2D seismic is generally used in early stages of exploration when no hydrocarbon reserves are discovered, yet [28].

3.3.1.1 Applications of 2D Seismic Data in Reservoir Characterization

Two-dimensional seismic data are analyzed slice-by-slice. Each slice is analyzed to interpret events such as horizons. Horizons can be distinguished as bright or dark

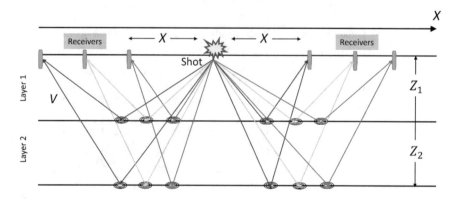

Fig. 3.4 Conceptualized configuration of the 2D seismic survey

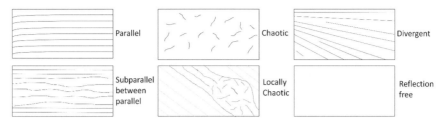

Fig. 3.5 Some examples of seismic textures (stratigraphic patterns) which are searched for during 2D seismic data interpretation

lines in gray-level reflection data [28]. Stratigraphic interpretation of the data is another feature that needs a proper interpretation alongside the horizons. The root of stratigraphy is the word stratum, which means a layer of rock with distinguishable characteristic from the adjacent layers. Areas with specific textures, or multiple horizons with similar trends construct the strata. This is known as seismic texture [29]. Figure 3.5 provides some examples of seismic textures/patterns. These patterns carry on very useful information. There are nearly a hundred of specific seismic textures/patterns, each of which representing a different rock-formation characteristics, also referred as lithologies [28]. Similar patterns represent lithologies from the same subgroup. Accordingly, even if the geologist cannot identify the precise lithological subgroup of a pattern/texture, he/she can distinguish its general group.

3.3.1.2 Limitations of 2D Seismic Surveys

- Only a thin section of the underground structures can be captured by 2D seismic data.
- Off-line reflections bring about some issues. These reflections sometimes refer to sideswipe and cannot easily be distinguished from the reflections directly below.

Sideswipe occurs in 2D seismic data where a feature appears out of the seismic plane, for example, fault, anticline, or other structures. Sideswipes are not seen in properly migrated 3D seismic surveys [30].

3.3.2 Three-dimensional Seismic Data

Three-dimensional seismic surveys are acquired from several lines of shot and receivers which are placed close to each other with tens of meters distance between them. Therefore, a 3D seismic survey creates a volume that is obtained by conducting multiple 2D surveys within a grid of several lines and interpolating between the lines to construct a 3D volume of seismic data, which is also known as a "cube" (see Figure 3.23 in [30]). Both onshore and offshore 3D surveys are obtained with

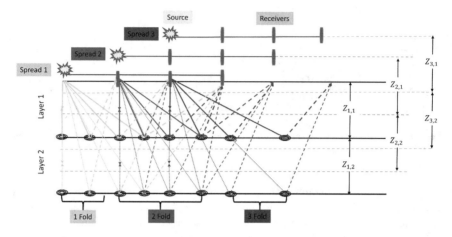

Fig. 3.6 Conceptualized configuration of the 3D seismic survey

several shot and receiver lines to assist the gathering of large volumes of data. The conceptualized configuration of 3D seismic is illustrated in Fig. 3.6.

3.3.2.1 Applications of 3D Seismic Data in Reservoir Characterization

Three-dimensional seismic surveys are often used in areas where hydrocarbons are already known to be present and where greater resolution and better-imaging of sub-surface geological features is needed. Seismic data are potent to establish a connection between core analysis results and wells, and well-test analysis results and tracer [24]. So far, the primary goal of geological maps and models constructed using 3D seismic data has been hydrocarbon exploration rather than production. Consequently, the external configuration and geometry of the reservoir has been the focus of 3D seismic interpretations rather than internal structure of the reservoir that controls the production mechanism. To have a better exploitation of the reservoir, it is required to have an adequate model that well describes the reservoir heterogeneity—vertical zonation, lateral compartmentalization, reservoir boundaries, hydrocarbon reserves, and the source of anisotropy and directionality of fluid flow in the reservoir.

3D seismic data can be analyzed to infer the reservoir geometry, adjust rock properties, and flow/production surveillance. The production strategy is envisaged based on this information. Direct measurements of petrophysical data at well locations are used to propagate the property of interest throughout the reservoir by an interpolation technique as described in Chap. 2. Nonetheless, the distribution of well information is uneven and sparse. If the reservoir is highly heterogeneous and complex, the model built based on these sparse data can result in tremendous errors and detrimental consequences on field development plans. Table 3.2 provides the list of information that can be acquired by analyzing seismic surveys and the related geological interpretation from the analysis [24].

Table 3.2 Seismic parameters and their stratigraphic interpretations [24]

Seismic parameters	Geological interpretation
Reflection configuration	Bedding patterns
	Depositional process
	Erosion and paleotography
	Fluid contacts
Reflection continuity	Bedding continuity
	Deposition process
Reflection amplitude	Velocity-density contrast
	Bed spacing
	Fluid content
Reflection frequency	Bed thickness
	Fluid content
Interval velocity	Estimate of lithology
	Estimate of porosity
	Fluid content
External form and areal association of seismic facies units	Cross sectional environment
	Sediment source
	Geologic setting

According to Table 3.2, 3D seismic surveys are used for constructing the static reservoir model as well as to estimate the hydrocarbon reserve of the sub-surface reservoir. Another application of 3D seismic data is that the spatially dense seismic attributes (e.g., acoustic impedance) can be incorporated to the sparse hard data by co-kriging for seismic inversion to build the prior geological realizations as the starting point of history matching. In this manner, the seismic data are used as soft/secondary data.

3.3.2.2 Advantages of 3D Seismic Surveys

- Density of data is often improved by 3D seismic surveys.
- 3D seismic addresses the issues of 2D seismic, such as sideswipe and out-of-plane reflections.
- Distribution of faults and the structure of underground formation can be realized by detailed information acquired from 3D seismic.
- The seismic cube provided by 3D surveys is a representation of a volume of the earth, which gives us the opportunity to interpret the data in various ways. The results of interpretation are regarded as an enhanced comprehension of structures and nature of sub-surface formations and improved probability of discovering recoverable hydrocarbon reserves.

3.3.2.3 Limitations of 3D Surveys

- 3D seismic is more expensive, involving more hardware on the ground and more computer processing power to process the large amount of data being collected.

3.3.3 Four-dimensional Seismic Data

4D seismic surveys are actually time lapses of 3D surveys in which 3D seismic data are acquired periodically over the same area. The purpose of 4D surveys is to monitor the changes in the pressure, saturation, temperature, and movement of the fluid in a producing petroleum reservoir with time [30]. 4D seismic data are used for surveillance of the way fluids flow within a reservoir, by conducting a baseline (before production begins) 3D survey (Fig. 3.7a) and repeating the survey several times during the production time of the reservoir (Fig. 3.7b). The repetition of 3D seismic surveys is known as 4D seismic survey.

To process 4D seismic data, the acquired data from the baseline surveys are subtracted from the acquired data from the monitor survey. Both baseline and monitor surveys are 3D. The difference between the two surveys is an indication of the changes in the reservoir. If the difference is zero, it means that no change has occurred in the reservoir.

3.3.3.1 Applications of 4D Seismic Data in Reservoir Characterization

One of the issues in production history matching (PHM) is that the production data are available only at well locations. 4D seismic data can be used to assess the fluid movement between well locations. In this regard, reservoir engineering can take

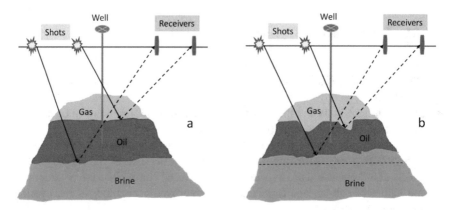

Fig. 3.7 a Conceptualization of the baseline 3D survey, and **b** of the monitor 3D survey (after production)

benefit from 4D seismic data to identify regions of the reservoir where pressure or fluid content are changed. Thereby, qualitative comparison between the measured and model-based predictions of 4D seismic data is more conventional than history matching, which can discover the problems with the model if done properly [31]. Qualitative comparison has a main shortcoming that cannot fully use the information content of the data. Therefore, it is not possible to efficiently assess the range of uncertainty in the forecasts [32].

Quantitative use of 4D surveys to update reservoir models has its own challenges, including the complicated process of substantially assimilating large amounts of data into prior models while honoring a large number of constraints, determining the weights of different kinds of data for simultaneous data assimilation, the difficulty of selecting parameters that can match data and keep admissible distribution of rock properties, the narrow uncertainty range in the prior models, noticeable errors in forward modeling of attributes and seismic data, and the extremely non-linear correlation between data and model parameters. The general workflow for production and 4D seismic history matching is illustrated in Fig. 3.8.

One point that worth of mentioning is the levels of seismic data integration in history matching. To do so, it is necessary to define a common domain where the simulation results and seismic data can be compared. Typically, there are various levels and forms that 4D seismic measurements can be represented. For example, 4D seismic observations can be represented in time, or depth, as seismic amplitudes, impedances, or even as pressure and maps of fluid saturation [32]. Totally, there are three main levels for comparison: (1) the seismic level, (2) the elastic parameter level, (3) the simulation model level. For more details please refer to [32].

3.3.3.2   Advantages of 4D Seismic Surveys

- Reservoir engineers and geoscientists can take benefit from 4D seismic and optimize static and dynamic models of complex reservoirs using state-of-the-art hardware. The outcome of such process will be enhanced oil recovery from fields, optimal locating of wells, decreased drilling costs and prolonged life of the field.
- Oil and gas fields can be better managed using 4D seismic surveys. For example, the expansion of an existing gas cap can be monitored to prevent the gas from reaching the production wells, which can subsequently reduce the oil production rate by increasing the gas-oil ratio.
- A baseline can be provided for development of other fields identified in the vicinity of the current field.
- Bypassed oil and gas can be discovered by 4D seismic surveys.

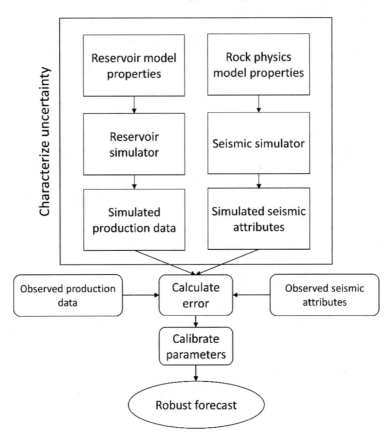

Fig. 3.8 Generic steps of seismic history matching (SHM)

3.3.4 Application of Static and Production Data

Static data are used as input data into geostatistical methods to generate geological realizations of the property of interest. Then, dynamic data are integrated to static data to achieve reliable reservoir models. As stated in earlier chapters, it is not a trivial task to construct a reservoir model since it is an inverse problem. Inverse problems are ill-posed problems that multiple solutions can be found to. This non-uniqueness of the solution is addressed by integrating as much data as possible to the problem. The static data are obtained from prior knowledge of the field to construct prior geological realizations of the sub-surface reservoir. Methods of creating geological realizations and utilizing the prior information are described in Chap. 2.

The remaining uncertainty after generating the prior realizations is reduced by integrating the dynamic data. Integration of production data in known as production history matching (PHM) and integration of seismic data in known as seismic history matching (SHM). Both PHM and SHM deal with minimizing the difference between the observed and simulated dynamic data by adjusting the static model properties using an optimizer. SHM uses 4D (also referred to as time-lapse 3D seismic) as dynamic data in the data assimilation step. The main steps of SHM are explained in Sect. 3.3.3. Further details can be found in [32, 33].

PHM involves assimilation of production data (production/injection rates, pressure, water-cut, etc.) to update model parameters with the purpose of establishing a match between the simulated and measured production data under a fixed production strategy honoring some constraints, such as preserving the geologic realism of the prior models (Eq. 3.2). The details of dynamic data assimilation and methods of production history matching are presented in Sect. 3.6. Note that, the mathematical details of seismic history matching are beyond the scope of this book, and all content about history matching refers to production history matching in the remainder of this book.

3.4 Challenges in History Matching

As Eq. 3.2 shows, history matching updates a vector of parameters that represents the uncertain properties of the reservoir model. However, in real-life, reservoir models are not represented in the vector form and have 2D or 3D structures with complex and mostly non-Gaussian distributions. Therefore, there are some challenges related to history matching which are introduced in the following.

3.4.1 High-dimensional Model Parameters

One of the main challenges in history matching is the high dimension of the model parameters vector, m. Since history matching is an optimization problem, as the dimension of m gets larger, the number of decision variables increases and makes the optimization problem more challenging. In this regard, if the uncertainty is in micro-scale parameters, such as permeability or porosity distribution, there will be a large number of optimization variables. This means that we will need more iterations of model updates to get a satisfactory match between the actual and simulated production data as well as a minimum deviation from the geologic realism of the prior model.

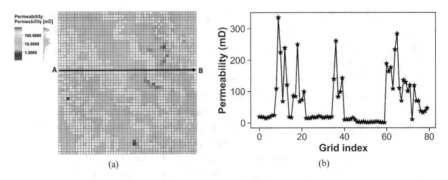

Fig. 3.9 Permeability distribution of a channelized reservoir model, **a** permeability map, **b** permeability variation between points A and B

3.4.2 Non-Gaussian and Non-linear Distribution of Reservoir Parameters

Another challenge in history matching is the non-Gaussian and non-linear distribution of reservoir properties. This is of a greater concern when dealing with microscale properties, such as permeability, porosity, or net-to-gross ratio (NTG). Since ensemble-based history matching methods assume that the input (uncertain reservoir properties) has a Gaussian distribution, their performance deteriorates if the reservoir properties have a non-Gaussian distribution [34]. Also, the non-linear structure of the model makes the updating rule more challenging. Figure 1-1 shows the non-Gaussian distribution of the permeability field of a channelized reservoir model, and Fig. 3.9 illustrates its corresponding variation of permeability on a straight line from point A to B on the permeability map of that model. As can be seen from Fig. 3.9b, the permeability has a non-linear trend from point A to B. Therefore, the updating rule has to capture these non-linear variations during the history matching process.

3.4.3 Three-dimensional Distribution of Model Parameters

Reservoir models consist of several geological layers deposited during different geological periods. Therefore, each of them has a different spatial distribution, which has deposited in different time. This means that, those layers deposited during an identical geological period follow the same spatial distribution. Nonetheless, this spatial distribution differs through time. As a result, we need to capture the distribution of spatial properties in the third dimension to have a better update of the properties in the course of history matching. Fig. 3.10 shows three random layers of five geological layers of a channelized reservoir model. As Fig. 3.10 shows, the spatial distribution of permeability in each geological layer is different due to variations in the orientation of channels from one layer to another.

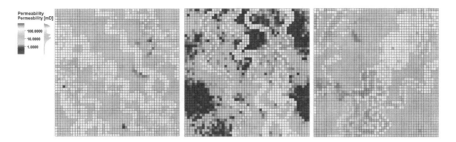

Fig. 3.10 Permeability distribution in three random layers of a channelized reservoir model

In Sect. 3.5, the common approaches to reservoir management under geological uncertainty, including the open-loop and closed-loop reservoir management are presented, and their advantages and disadvantages are discussed.

3.5   Different Approaches to Reservoir Management Under Geological Uncertainty

Uncertainty management using observed production data entails a step of conditioning the prior reservoir model realizations to historical production data. In this regard, the conditioning can be done on a single realization or on an ensemble of realizations as explained in Sect. 3.1. There are different approaches to conditioning the realizations to production data each of which having its own pros and cons. Generally, there are two main types of approaches to reservoir management under geological uncertainty, including the open-loop and closed-loop reservoir management. In the following, these approaches are described and their pros cons are discussed.

3.5.1   Open-Loop Reservoir Management (OLRM)

The conventional approach used in the industry is to adjust the uncertain parameters of the model manually to get a match between the historical simulated and measured production data [23]. The manual approach is time-consuming and requires intensive human labor. In the meantime, the model conditioning (history matching) phase can be done only once, and the obtained model(s) can be used for decision-making in all further field development activities for the rest of the reservoir's life. This process is known as open-loop reservoir management (OLRM). This approach is the most used reservoir management strategy in the industry and literature [35–39]. The overall steps of OLRM are demonstrated in Fig. 3.11. According to Fig. 3.11, OLRM entails two main steps: history matching and field development optimization. In the history matching step, a reliable reservoir model, or an ensemble of reservoir models are

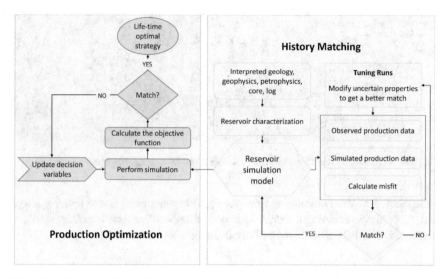

Fig. 3.11 Main steps of the OLRM

obtained, and then used for optimization of field development variables. The reservoir updating step is done only once and any field development optimization is conducted using the results of the history matching step.

The advantage of open-loop reservoir management is that it is computationally less expensive since data assimilation is done only once. Also, uncertainty quantification is done only once at the time of creating prior realizations. Nonetheless, uncertainty cannot be reduced significantly using the limited data at early stages of field development. As time passes, more static, seismic, and production data become available that can falsify our prior knowledge of the reservoir and show deviation from the predicted reservoir behavior.

3.5.2 Closed-Loop Reservoir Management (CLRM)

Our knowledge from newly discovered reservoirs is limited. As time goes on and we get more production data from the reservoir, we can update the reservoir model based on new data to further reduce the uncertainty in reservoir characterization. Now, having a better reservoir model(s), field development optimization can be continued using the new reservoir model(s). This practice is repeated periodically as more data are obtained from the reservoir. This process is known as the Closed-Loop Reservoir Management (CLRM), also known as real-time reservoir management [14, 40–44]. Figure 3.12 shows the main steps of the CLRM [40].

Typically, if the data assimilation and production optimization (field development optimization) is repeated for at least two cycles, it is called closed-loop. Generally,

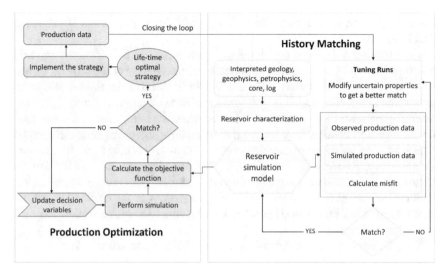

Fig. 3.12 Main steps of the CLRM

closed-loop is described as a two-component process, consisting of a data assimilation step followed by a production optimization step during each cycle [14, 45]. In other works, CLRM is introduced as a three-component process, including the optimization, model updating, and uncertainty propagation [46]. In a recent work by Mirzaei-Paiaman et al., four components are considered for closed-loop, in which the following actions are included in each cycle [23].

1. Measurement (or surveillance): gathering new data (for example, production/injection rates/volumes, well logs, and seismic surveys);
2. Data assimilation: modifying uncertain reservoir/field properties;
3. Production optimization and decision-making: conducting an optimization to find an optimum production strategy (e.g., type, number, position of the infill wells, operational conditions of existing and new wells)
4. Implementation (or operation/execution): implementing the selected production strategy in the field/reservoir (for example, drilling new wells, and adjusting well controls).

During the history matching phase, a model or an ensemble of models is updated to minimize the misfit function (e.g., Eqs. 3.2 or 3.3). The ensemble of models is also referred to as the fit-for-purpose models [23]. These models are the upscaled versions of the detailed and fine-grid geological models [47]. One can trade-off between the fidelity of the fit-for-purpose models and the available computational hardware and time. Furthermore, another approach is to construct multi-fidelity models to take benefit from both high precision models and a low computational cost [48]. It is worth mentioning that if the production optimization step involves field-scale variables, the whole process is called the Closed-Loop Field Development and Management (CLFDM) [23].

Throughout the life span of the field, as new information is obtained from the reservoir, the models are updated to decrease the uncertainty of the models. Consequently, more reliable models will be available to conduct the field development optimization step. Although it is mentioned in almost all previous works that as this cyclic process advances we get more information on the field, which are assimilated to the models to continuously reduce the uncertainty, there can be some complex cases where acquiring more knowledge about the reservoir may result in increased uncertainty from one cycle to another [23]. For example, new data can indicate the existence of a fracture network which was not inferred from previous information. Different authors have investigated the implementations of the CLRM [13, 14, 16, 20, 40, 45, 49] and the CLFDM [18, 23, 50, 51], which interested readers can refer to for more information.

Now, different history matching (data assimilation) methods are described and their pros and cos are discussed.

3.6 History Matching Methods

History matching methods refer to the sets of mathematical formulations and systematic approaches which are used for data assimilation to update uncertain reservoir properties. These methods are described theoretically and mathematically in the following, and their benefits and limitations are explained.

3.6.1 Ensemble-based Methods

Ensemble-based history matching methods refer to those methods that update an ensemble of realizations during the data assimilation process. For decades, ensemble-based methods have been extensively used for data assimilation within problems regarding flow related to atmospheric physics and hydrocarbon reservoir history matching. Among ensemble-based history matching methods, Ensemble Kalman Filter (EnKF) and Ensemble Randomized Maximum Likelihood (EnRML) are two of the most used methods in a wide range of problems.

3.6.1.1 Ensemble Kalman Filter (EnKF)

Traditional history matching methods do not provide any facility to continuously update the model, but they make use of all assimilated data at the same time to construct a reservoir model. To address some of the shortcomings of traditional methods, methods with sequential data assimilation are utilized. Ensemble Kalman

Filter (EnKF) is one of the most used methods in hydrocarbon reservoir history matching and management [13, 15, 49, 52, 53]. This method was first introduced by Evensen [54] and modified by Burgers et al. [55] is a Monte Carlo (MC) approximation of the Kalman filter [56] that uses Monte Carlo sampling to forecast the error statistics. The first application of the EnKF in reservoir history matching was by Naevdal et al. [57] where they used the EnKF to update the permeability field for near-well reservoir models. The data assimilation process in EnKF is sequential, which means that the assimilation occurs in different time-steps throughout the reservoir's life-span. This characteristic of the EnKF makes it suitable for CLRM. In this method an ensemble of parameter-state vectors is integrated forward in time, and as new data is assimilated, the ensemble is updated. In this method, the parameter-state vector consists of unknown model parameters of the reservoir, such as the petrophysical and geological properties, as well as the vector of dynamic state properties (e.g., saturation, pressure, and dissolved gas-oil ratio of gridblocks in a conventional black-oil reservoir simulator). The reason for including dynamic state of the reservoir in the data assimilation process is to mitigate the need for running the simulations from time zero at each data assimilation time-step [58]. Nevertheless, it is required to assume a statistical consistency between the updated vectors of states and model parameters. The statistical consistency is described in [59]. The mathematical formation and assumption of the EnKF are presented in the following.

Let $y^n = \begin{bmatrix} m^n \\ p^n \end{bmatrix}$ be the N_y-dimensional parameter-state vector at the nth data assimilation time-step where m^n denotes the N_m-dimensional vector of model parameters, and p^n represent the N_p-dimensional vector of dynamic parameters. p^n includes dynamic response of the reservoir (e.g., pressure, saturations, etc.). Accordingly, the formulation of the EnKF is given as follows:

$$y_j^{n,a} = y_j^{n,f} + C_{YD}^{n,f} \left(C_{DD}^{n,f} + C_D^n \right)^{-1} \left(d_{uc,j}^{n,f} - d_j^{n,f} \right), \quad j = 1, 2, \ldots, N_e. \quad (3.4)$$

In Eq. 3.4, $y^{n,a}$ and $y^{n,f}$ are the *analysis* and *forecast* parameter-state vectors at the nth data assimilation time-step, $C_{DD}^{n,f}$ is the auto-covariance of the simulated data of size $N_n \times N_n$, C_D^n is the covariance matrix of measurement errors with a size of $N_n \times N_n$, $C_{YD}^{n,f}$ is the covariance matrix between the *forecast* parameter-state vector and the simulated data of size $N_y \times N_n$, d_{uc}^n represents the N_n-dimensional vector of perturbed observation data with a Gaussian noise (i.e., $d_{uc}^n \sim \mathcal{N}(d_{obs}^n, C_D^n)$) in which d_{obs}^n is the N_n-dimensional vector of observed production data at the nth data assimilation time-step, $d^{n,f}$ denotes the vector of predicted data by the reservoir simulator at the nth data assimilation time-step, N_e is the ensemble size, and N_n represents the number of assimilated data points at the nth time-step.

Although EnKF has been widely used in history matching problems, it has some drawbacks that limit its applications. Zafari and Rynolds [60] found that the EnKF is not able to handle multimodal non-linear problems (non-Gaussian distributions). This is because the standard EnKF assumes that the parameters-state vector has a Gaussian distribution. Therefore, for input data such as facies distributions that have a bimodal distribution, the EnKF cannot be applied since it cannot preserve the geologic realism [53]. Moreover, since the EnKF updates the model one-time-step by one-time-step, it can lead to inconsistency between the updated static model parameters and the dynamic state of the model [58]. Another shortcoming of the EnKF is that it requires many simulation restarts that can extremely slow down the history matching process if the reservoir model is large. Two other limitations of the EnKF that are reported in the literature are (1) finite size of the ensemble and, (2) inappropriate initial ensemble [61].

To overcome the issues related to non-Gaussian distribution of the model parameters, many authors proposed an iterative form of the EnKF [62–65]. Some other researchers have proposed to use parameterization of the parameter-state vector to overcome the non-Gaussian effects [3, 15, 66, 67]. To resolve the effect of inappropriate initial ensemble, the authors in [68–70] suggested to use the iterative form of the EnKF. An iterative form of the EnKF is known as the Ensemble Randomized Maximum Likelihood (EnRML) that is described in the following.

3.6.1.2 Ensemble Randomized Maximum Likelihood (EnRML)

This method is an iterative form of the EnKF that was introduced by Gu and Oliver [65]. This method updates the model parameters using a Gauss-Newton update equation:

$$
\begin{aligned}
m_j^{\ell+1} = \beta_\ell m_j^f + (1 - \beta_\ell)m_j^\ell - \beta_\ell C_M^f \overline{G}_\ell^{\mathrm{T}} \left(\overline{G}_\ell C_M^f \overline{G}_\ell^{\mathrm{T}} + C_D \right)^{-1} \\
\times \left[d_j^\ell - d_{\mathrm{uc},j} - \overline{G}_\ell \left(m_j^\ell - m_j^f \right) \right],
\end{aligned} \tag{3.5}
$$

for $j = 1, 2, \ldots, N_e$. In Eq. 3.5, ℓ indicates the number of iterations and β_ℓ is the step-size. C_M^f remains fixed throughout the iterative process and is estimated according to the forecast ensemble. $\overline{G}$ denotes the average sensitivity matrix that is computed according to the ensemble by solving $\Delta D^\ell = \overline{G}_\ell \Delta M^\ell$ for $\overline{G}_\ell$ using singular value decomposition (SVD) [71]. ΔD^ℓ can be calculated as $D^\ell - \overline{D}^\ell$, in which D^ℓ represents the matrix of the ensemble of predicted data at the ℓth iteration. In other words, predictions on the jth ensemble member is placed at the jth column of D^ℓ. $\overline{D}^\ell$ is a matrix that all its columns are $\overline{d}^\ell$ indicating the average of all comuns of D^ℓ. In a similar manner, $\Delta M^\ell = M^\ell - \overline{M}^\ell$ where each column of M^ℓ represents a vector of model parameters of the jth member of the ensemble at the ℓth iteration. Similarly, the matrix with all columns equal to $\overline{m}^\ell$ representing the average of all columns of M^ℓ is denoted by $\overline{M}^\ell$. If the model updated by Eq. 3.5 does not improve the match between the predicted and observed data, a line search has to be performed.

EnRML has been used by different authors for history matching of petroleum reservoirs. Chen and Oliver [72] combined the Levenberg-Marquardt with the EnRML (LM-EnRML) to history match the NORNE field using production data and RFT pressures. They also suggested to use a distance-based localization to regularize the updates. For more information on the LM-EnRML and localization please refer to [72]. Also, Evensen et al. [73] proposed an extension to the EnRML, known as the subspace-EnRML for data assimilation and history matching of reservoirs.

Although EnRML-based methods have been widely used in history matching problems, they have not shown satisfactory results and entail large computational costs as an iterative EnKF method [58].

3.6.2 Ensemble-Smoother Methods

Petroleum reservoirs show a more stable behavior compared to atmospheric and oceanic models [74]. As a consequence, it might not be necessary to update the model recursively using methods such as the EnKF and EnRML since these methods have demanding computational cost. Accordingly, ensemble-smoother-based methods that assimilate all historical data and perform a model update are proposed in the recent years. These methods were proposed to resolve the shortcomings of the EnKF and EnRML, such as the inconsistency between the updated model state and model parameters. Ensemble Smoother and Ensemble Smoother—Multiple Data Assimilation are two most used ensemble-smoother-based methods which are described in the following.

3.6.2.1 Ensemble Smoother (ES)

Equation 3.6 gives the formulation of the ensemble smoother to get the posterior model parameters based on the prior model parameters and the assimilated production data:

$$m_j^{posterior} = m_j^{prior} + C_{MD}^{prior}(C_{DD} + C_D)^{-1}(d_{uc,j} - d_j), \quad j = 1, 2, \ldots, N_e. \tag{3.6}$$

where m^{prior} and $m^{posterior}$ are the vectors of prior and posterior model parameters. C_{MD}^{prior} is the matrix of cross-covariance between the prior model parameters and the vector of simulated data, C_{DD} is an $N_d \times N_d$ matrix of auto-covariance of the simulated data, C_D is an $N_d \times N_d$ matrix of covariance of data measurement errors, $d_{uc}^n \sim \mathcal{N}(d_{obs}, C_D)$ is the perturbed observed data as in the EnKF, and d represents the vector of the simulated production data.

Ensemble Smoother has been used by different authors for history matching of model parameters. Lee et al. [75] compared the ES with EnKF and concluded that the computational cost of the ES is about 2.2% of the EnKF. Liao et al. [76] also employed the ES on a 2D synthetic reservoir and concluded that the performance of the ES worsens as the heterogeneity (i.e., non-linearity) increases in the model parameters. As a consequence, to handle the non-linearity in the input parameters, an iterative form of the ensemble smoother was proposed.

3.6.2.2 Ensemble Smoother—Multiple Data Assimilation (ES-MDA)

Ensemble Smoother with Multiple Data Assimilation is an ensemble-based history matching method in which an ensemble of prior model realizations is conditioned to historical production data through an iterative approach with multiple data assimilations. The formulation of the ES-MDA is expressed as:

$$m_j^{t+1} = m_j^t + C_{md}^t \left(C_{dd}^t + \alpha_t C_d \right)^{-1} \left(d_{obs} + e_j^t - d_j^t \right), \qquad (3.7)$$

where m_j^{t+1} and m_j^t are the jth realizations at iteration $t + 1$ and t, respectively, $C_{md}^t \in \mathbb{R}^{N_m \times N_d}$ is the cross-covariance matrix between the model parameters and observed data, and $C_{dd}^t \in \mathbb{R}^{N_d \times N_d}$ is the covariance matrix of the predicted data at iteration t. $C_d \in \mathbb{R}^{N_d \times N_d}$ denotes the data-error covariance matrix, α_t is the inflation coefficient $\left(\sum_{k=1}^{N_a} \alpha_k^{-1} = 1 \right)$, N_a is the number of data assimilation steps, $d_{obs} \in \mathbb{R}^{N_d}$ is the vector of observed data, $d_j^t \in \mathbb{R}^{N_d}$ is the vector of predicted data, and $e_j^t \in \mathbb{R}^{N_d}$ is the vector of random error which is defined as: $e_j^t \sim \mathcal{N}(0, \alpha_t C_d)$.

Emerick and Reynolds [74] provided two procedures to prove that the single and multiple data assimilations are equivalent for linear-Gaussian cases. In the first procedure, using the Kalman filter equations, they showed that the mean and covariance obtained from updating by MDA are equivalent to the posterior mean and covariance that are obtained by a single data assimilation using the actual covariance of the measurement errors, C_d. In the second procedure, they showed that, if a sample is taken from the prior probability density function, after assimilating data with MDA, it is from the correct posterior PDF. In spite of the equivalence between the two procedures, the second procedure shows the necessity of perturbing the observations using the inflated C_d to get the right posterior distribution for the linear-Gaussian case. The generalization of the second proof can be found in [58].

However, that proof holds for the linear-Gaussian case. The reason for applying multiple data assimilation for nonlinear cases is that the ensemble smoother is equivalent to a single Gauss-Newton iteration with a full step and an average sensitivity estimated from the prior realizations [77]. In this manner, ES-MDA can be regarded as an iterative form of the ES, where we update the models using multiple smaller corrections instead of a single and large Gauss-Newton update. Note that the number of iterations in ES-MDA is determined beforehand. The workflow of ES-MDA is described in the following [58]:

1. Determine the number of steps for data assimilation, N_a, and the inflation coefficients, α_i, for $i = 1, 2, \ldots, N_a$.
2. For $i = 1$ to N_a:

   a. Simulate the ensemble from time zero to the end of the simulation time.

    b. For each ensemble member, add a Gaussian noise to the vector of predicted data, $d_{uc} = d_{obs} + \sqrt{\alpha_i} C_d^{1/2} z_d$, where $z_d \sim \mathcal{N}(0, I_{N_d})$.

    c. Use Eq. 3.7 and update the ensemble.

3.6.3 Stochastic Optimization Algorithms

Stochastic optimization algorithms do not use the covariance of models parameters and production data, instead, they directly update the model parameters to minimize the misfit function. These algorithms include the swarm intelligence and evolutionary algorithms.

3.6.3.1 Evolutionary Algorithms

Evolutionary algorithms are meta-heuristic algorithms based on the Darwin's natural selection law. Genetic Algorithm (GA), Simulated Annealing, Genetic Programming, Evolutionary Programming, Differential Evolution, and Neuroevolution are some examples of evolutionary algorithms. In this regard, GA is one of the most popular evolutionary algorithms that was first introduced by Goldeberg [78]. This algorithm starts with a random generation and searches for the global optimum of the objective function by updating the optimization variables using three operations, including the selection, cross-over, and mutation. In the context of history matching, the optimization variables are the model parameters (e.g., permeability of all gridblocks, depth of water/oil contact, etc.). In this procedure, prior realizations are simulated and production data are obtained. Then, the misfit function is calculated and the realizations are updated using the three operations of the GA. These steps are repeated until there is no significant change in the misfit function from one generation to the next.

    Different authors have implemented the GA for history matching of petroleum reservoirs. For example, Xavier et al. [79] employed the GA for automatic history matching of a two-dimensional two-phase reservoir model. Lange [80] coupled the GA with a discrete fracture network (DFN) flow simulator in a history matching problem with the purpose of characterizing fractured reservoir models in consistency with the well-test data. Ballester and Carter [81] applied a real-coded GA to history matching of a real reservoir model and showed the applicability of this algorithm to complex and real reservoir models. Another example is the application of an extension to the GA with adaptive mutation and crossover operations to reservoir history matching that showed improved performance upon the standard GA [82]. Other examples can be found in [83–85]. For more details on the genetic algorithm please refer to [78, 82, 86, 87].

3.6.3.2 Swarm Intelligence (SI) Algorithms

Swarm Intelligence algorithms are a part of evolutionary algorithms inspired by the collective behavior of animals, searching for food or survival [88]. Swarm intelligence algorithms contain a wide variety of algorithms, such as the Ant-Colony, Bees, Shuffled Frog Leap, Firefly, Glow-worm, Lion, Wolf, Bat, Monkey, and Particle Swarm Optimization (PSO) algorithms. These algorithms have had various applications that can be found in [88]. Among other, PSO has gained more prominence in petroleum engineering applications. PSO is a population-based optimization algorithm that was first introduced by Kennedy and Eberhart in 1995 [89]. In this algorithm, each solution (model parameters) is represented by a particle, where a group of particles forms the swarm. This algorithm uses the best solutions obtained by the swarm and individual particles to update the solution. Interested readers can refer to [89, 90] for more details on the PSO.

PSO has been widely used for history matching of reservoir model parameters by different authors. Sanghyun and Stephen [83] applied both the GA and PSO to history matching of an ensemble of realizations on a large reservoir model and found that using evolutionary algorithms is suitable for updating a large number of matching parameters. Awotunde [91] proposed an improved PSO that used an averaging filter to decrease the number of function evaluations in history matching of reservoir model parameters. Zhang et al. [92] combined the PSO and the gravitational search algorithm (PSO-GSA) to history match petroleum reservoirs and concluded that the proposed algorithm outperformed the standard PSO and the genetic algorithm. Other examples can be found in [93–97].

References

1. Oliver DS, Chen Y (2010) Recent progress on reservoir history matching: a review. Compu Geosci 15(1):185–221. https://doi.org/10.1007/S10596-010-9194-2
2. Ghoniem SA, Abdel Aliem S, Sayyouh MH, El Salaly M (1984) A simplified method for petroleum reservoir history matching. Appl Math Model 8(4):282–287. https://doi.org/10.1016/0307-904X(84)90163-X
3. Heidari L, Gervais V, Le Ravalec M, Wackernagel H (2013) History matching of petroleum reservoir models by the ensemble Kalman filter and parameterization methods. Comput Geosci 55:84–95. https://doi.org/10.1016/j.cageo.2012.06.006
4. Zhang Y, Oliver DS (2011) History matching using the ensemble Kalman filter with multiscale parameterization: a field case study. SPE J 16(02):307–317. https://doi.org/10.2118/118879-PA
5. Hajizadeh Y, Christie M, Demyanov V (2011) Ant colony optimization for history matching and uncertainty quantification of reservoir models. J Petrol Sci Eng 77(1):78–92. https://doi.org/10.1016/j.petrol.2011.02.005
6. Subbey S, Christie M, Sambridge M (2004) Prediction under uncertainty in reservoir modeling. J Petrol Sci Eng 44(1–2):143–153. https://doi.org/10.1016/j.petrol.2004.02.011
7. Ghaedi M, Masihi M, Heinemann ZE (2014) History matching of naturally fractured reservoirs based on the recovery curve method. J Petrol Sci Eng 1–11. https://doi.org/10.1016/j.petrol.2014.12.002

8. Peters E, Arts RJ, Brouwer GK, Geel CR, Cullick S, Lorentzen RJ, Chen Y, Dunlop KNB, Vossepoel FC, Xu R et al (2010) Results of the brugge benchmark study for flooding optimization and history matching. SPE Reservoir Eval Eng 13(03):391–405. https://doi.org/10.2118/119094-PA

9. Bukshtynov V, Volkov O, Durlofsky LJ, Aziz K (2015) Comprehensive framework for gradient-based optimization in closed-loop reservoir management. Comput Geosci 19(4):877–897. https://doi.org/10.1007/s10596-015-9496-5

10. Overbeek KM, Brouwer DR, Neavdal G, Kruijsdijk CPJW van, Jansen JD (2004) Closed-loop waterflooding. Earth Doc cp-9-00043. https://doi.org/10.3997/2214-4609-PDB.9.B033

11. Brouwer DR, Nævdal G, Jansen JD, Vefring EH, Van Kruijsdijk CPJW (2004) Improved reservoir management through optimal control and continuous model updating. Proc—SPE Ann Tech Conf Exhibit 26:1551–1561. https://doi.org/10.2118/90149-MS

12. Aitokhuehi I, Durlofsky LJ (2005) Optimizing the performance of smart wells in complex reservoirs using continuously updated geological models. J Petrol Sci Eng 48(3–4):254–264. https://doi.org/10.1016/j.petrol.2005.06.004

13. Nævdal G, Brouwer DR, Jansen JD (2006) Waterflooding using closed-loop control. Comput Geosci 10(1):37–60. https://doi.org/10.1007/S10596-005-9010-6

14. Jansen JD, Douma SD, Brouwer DR, Van Den Hof PMJ, Bosgra OH, Heemink AW (2009) Closed-loop reservoir management. SPE Reservoir Simul Sympos Proc 2(1):856–873. https://doi.org/10.3997/1365-2397.2005002/CITE/REFWORKS

15. Chen Y, Oliver DS, Zhang D (2009) Efficient ensemble-based closed-loop production optimization. SPE J 14(04):634–645. https://doi.org/10.2118/112873-PA

16. Wang C, Li G, Reynolds AC (2009) Production optimization in closed-loop reservoir management. SPE J 14(03):506–523. https://doi.org/10.2118/109805-PA

17. Hanea R, Evensen G, Hustoft L, Ek T, Chitu A, Wilschut F (2015) Reservoir management under geological uncertainty using fast model update. In: SPE reservoir simulation symposium. Houston, Texas

18. Shirangi MG, Durlofsky LJ (2015) Closed-loop field development optimization under uncertainty. In: SPE reservoir evaluation and engineering. Houston, pp 23–25. https://doi.org/10.2118/173219-MS

19. Sarma P, Durlofsky LJ, Aziz K, Chen WH (2006) Efficient real-time reservoir management using adjoint-based optimal control and model updating. Comput Geosci 10(1):3–36. https://doi.org/10.1007/s10596-005-9009-z

20. Hui Z, Yang L, Jun Y, Kai Z (2011) Theoretical research on reservoir closed-loop production management. Sci China Technol Sci 54(10):2815–2824. https://doi.org/10.1007/S11431-011-4465-2

21. Avansi GD, Schiozer DJ (2015) UNISIM-I: synthetic model for reservoir development and management applications. Int J Model Sim Petrol Ind 9(1):21–30

22. Morosov AL, Schiozer DJ (2017) Field-development process revealing uncertainty-assessment pitfalls. SPE Reservoir Eval Eng 20(03):765–778. https://doi.org/10.2118/180094-PA

23. Mirzaei-Paiaman A, Santos SMG, Schiozer DJ (2021) A review on closed-loop field development and management. J Petrol Sci Eng 201:108457. https://doi.org/10.1016/j.petrol.2021.108457

24. Aminzadeh F, Dasgupta SN (2013) Reservoir Characterization. https://doi.org/10.1016/B978-0-444-50662-7.00006-8

25. Dos Santos JMC, Davolio A, Schiozer DJ, Macbeth C (2018) Semiquantitative 4D seismic interpretation integrated with reservoir simulation: application to the Norne field. Interpretation 6(3):T601–T611. https://doi.org/10.1190/INT-2017-0122.1

26. Alfi M, Hosseini SA (2016) Integration of reservoir simulation, history matching, and 4D seismic for CO2-EOR and storage at Cranfield, Mississippi, USA. Fuel 175:116–128. https://doi.org/10.1016/j.fuel.2016.02.032

27. Bogan C, Johnson D, Litvak M, Stauber D (2003) Building reservoir models based on 4D seismic and well data in gulf of mexico oil fields. Proc—SPE Ann Tech Conf Exhibit 2723–2733. https://doi.org/10.2523/84370-ms

28. Patel D, Giertsen C, Thurmond J, Gjelberg J, Gröller ME (2008) The seismic analyzer: inter-
    preting and illustrating 2D seismic data. IEEE Trans Visual Comput Graph 14(6):1571–1578.
    https://doi.org/10.1109/TVCG.2008.170
29. Randen T, Sønneland L (2005) Atlas of 3D seismic attributes BT—mathematical methods
    and modelling in hydrocarbon exploration and production. In: Iske A, Randen T (eds). Berlin,
    Heidelberg: Springer Berlin Heidelberg, pp 23–46. https://doi.org/10.1007/3-540-26493-0_2
30. Onajite E (2014) Understanding seismic exploration. Seismic data analysis techniques in
    hydrocarbon exploration, 33–62. https://doi.org/10.1016/b978-0-12-420023-4.00003-4
31. Maleki M, Davolio A, Schiozer DJ (2018) Using simulation and production data to resolve
    ambiguity in interpreting 4D seismic inverted impedance in the Norne Field. Petrol Geosci
    24(3):335–347. https://doi.org/10.1144/petgeo2017-032
32. Oliver DS, Fossum K, Bhakta T, Sandø I, Nævdal G, Lorentzen RJ (2021) 4D seismic history
    matching. J Petrol Sci Eng 207(June):109119. https://doi.org/10.1016/j.petrol.2021.109119
33. Skjervheim JA, Evensen G, Aanonsen SI, Ruud BO, Johansen TA (2005) Incorporating 4D
    seismic data in reservoir simulation models using ensemble Kalman filter. In: Proceedings—
    SPE annual technical conference and exhibition, pp 1497–1506. https://doi.org/10.2118/957
    89-ms
34. Canchumuni SWA, Emerick AA, Pacheco MAC (2018) History matching channelized facies
    models using ensemble smoother with a deep learning parameterization. In: 16th European
    conference on the mathematics of oil recovery, ECMOR 2018. 2018;(September). https://doi.
    org/10.3997/2214-4609.201802277
35. Zhang K, Zhang J, Ma X, Yao C, Zhang L, Yang Y, Wang J, Yao J, Zhao H (202AD) of naturally
    fractured reservoirs using a deep sparse autoencoder. SPE J 26(04):1700–1721. https://doi.org/
    10.2118/205340-pa
36. Zhao Y, Forouzanfar F, Reynolds AC (2017) History matching of multi-facies channelized reser-
    voirs using ES-MDA with common basis DCT. Comput Geosci 21(5–6):1343–1364. https://
    doi.org/10.1007/s10596-016-9604-1
37. Alguliyev R, Aliguliyev R, Imamverdiyev Y, Sukhostat L (2022) History matching of petroleum
    reservoirs using deep neural networks. Intell Syst Appl 16(September):200128. https://doi.org/
    10.1016/j.iswa.2022.200128
38. Wang Z, He J, Milliken WJ, Wen XH (2021) Fast history matching and optimization using
    a novel physics-based data-driven model: an application to a diatomite reservoir. SPE J
    26(06):4089–4108. https://doi.org/10.2118/200772-PA
39. Lorentzen RJ, Luo X, Bhakta T, Valestrand R (2019) History matching the full norne field
    model using seismic and production data. SPE J 24(04):1452–1467. https://doi.org/10.2118/
    194205-PA
40. Hou J, Zhou K, Zhang XS, Kang XD, Xie H (2015) A review of closed-loop reser-
    voir management. Petrol Sci 12(1):114–128. https://doi.org/10.1007/S12182-014-0005-6/TAB
    LES/3
41. Khor CS, Elkamel A, Shah N (2017) Optimization methods for petroleum fields development
    and production systems: a review. Optim Eng 18(4):907–941. https://doi.org/10.1007/S11081-
    017-9365-2
42. Benndorf J, Jansen JD (2017) Recent developments in closed-loop approaches for real-time
    mining and petroleum extraction. Math Geosci 49(3):277–306. https://doi.org/10.1007/S11
    004-016-9664-8
43. Oliver DS, Reynolds AC, Liu N (2008) Inverse theory for petroleum reservoir characterization
    and history matching
44. Udy J, Hansen B, Maddux S, Petersen D, Heilner S, Stevens K, Lignell D, Hedengren JD
    (2017) Review of field development optimization of waterflooding, EOR, and well placement
    focusing on history matching and optimization algorithms. Processes 5(3):34. https://doi.org/
    10.3390/PR5030034
45. Chen C, Li G, Reynolds AC (2012) Robust constrained optimization of short- and long-term
    net present value for closed-loop reservoir management. SPE J 17(03):849–864. https://doi.
    org/10.2118/141314-PA

46. Sarma P, Durlofsky LJ, Aziz K (2008) Computational techniques for closed–loop reservoir modeling with application to a realistic reservoir. Petrol Sci Technol 26(10–11):1120–1140. https://doi.org/10.1080/10916460701829580

47. Schiozer DJ, De Souza Dos Santos AA, De Graça Santos SM, Von Hohendorff Filho JC (2019) Model-based decision analysis applied to petroleum field development and management. Oil and Gas Science and Technology—Revue d'IFP Energies nouvelles. 74:46. https://doi.org/10.2516/OGST/2019019

48. Thenon A, Gervais V, Le RM (2016) Multi-fidelity meta-modeling for reservoir engineering—application to history matching. Comput Geosci 20(6):1231–1250. https://doi.org/10.1007/s10596-016-9587-y

49. Lorentzen RJ, Shafieirad A, Naævdal G (2009) Closed loop reservoir management using the ensemble Kalman filter and sequential quadratic programming. SPE Reservoir Simul Sympos Proc 2:886–897. https://doi.org/10.2118/119101-MS

50. Shirangi MG (2019) Closed-loop field development with multipoint geostatistics and statistical performance assessment. J Comput Phys 390:249–264. https://doi.org/10.1016/J.JCP.2019.04.003

51. Jahandideh A, Jafarpour B (2018) Hedging against uncertain future development plans in closed-loop field development optimization. In: Proceedings—SPE annual technical conference and exhibition. 2018-Septe. https://doi.org/10.2118/191622-ms

52. Lorentzen RJ, Berg AM, Nævdal G, Vefring EH (2006) A new approach for dynamic optimization of water flooding problems. In: 2006 SPE intelligent energy conference and exhibition, 1:253–263. https://doi.org/10.2118/99690-MS

53. Gu Y, Oliver DS (2006) The ensemble Kalman filter for continuous updating of reservoir simulation models. J Energy Resour Technol 128(1):79–87. https://doi.org/10.1115/1.2134735

54. Evensen G (1994) Sequential data assimilation with a nonlinear quasi-geostrophic model using Monte Carlo methods to forecast error statistics. J Geophys Res: Oceans 99(C5):10143–10162. https://doi.org/10.1029/94JC00572

55. Burgers G, van Leeuwen PJ, Evensen G (1998) Analysis scheme in the ensemble Kalman filter. Monthly Weather Rev 126(6):1719–1724. https://doi.org/10.1175/1520-0493(1998)126

56. Kalman RE (1960) A new approach to linear filtering and prediction problems. J Basic Eng 82(1):35–45. https://doi.org/10.1115/1.3662552

57. Nævdal G, Mannseth T, Vefring EH (2002) Near-well reservoir monitoring through ensemble Kalman filter. Proc—SPE Sympos Improved Oil Recovery 1:959–967. https://doi.org/10.2118/75235-ms

58. Emerick AA, Reynolds AC (2012) Ensemble smoother with multiple data assimilation. Comput Geosci 55:3–15. https://doi.org/10.1016/j.cageo.2012.03.011

59. Thulin, K., Li G, Aanonsen SI, Reynolds AC (2007) Estimation of initial fluid contacts by assimilation of production data with EKF. In: Proceedings of the SPE Annual Technical Conference and Exhibition. Anaheim, CA

60. Akter F, Imtiaz S, Zendehboudi S, Hossain K (2021) Modified ensemble Kalman filter for reservoir parameter and state estimation in the presence of model uncertainty. J Petrol Sci Eng 199:108323. https://doi.org/10.1016/J.PETROL.2020.108323

61. Jung S, Lee K, Park C, Choe J (2018) Ensemble-based data assimilation in reservoir characterization: a review. Energies 11(2):445. https://doi.org/10.3390/EN11020445

62. Wang Y, Li M (2011) Reservoir history matching and inversion using an iterative ensemble Kalman filter with covariance localization. Petrol Sci 8(3):316–327. https://doi.org/10.1007/S12182-011-0148-7

63. Lorentzen RJ, Naevdal G (2011) An iterative ensemble kalman filter. IEEE Trans Autom Control 56(8):1990–1995. https://doi.org/10.1109/TAC.2011.2154430

64. Li G, Reynolds AC (2009) Iterative ensemble Kalman filters for data assimilation. SPE J 14(03):496–505. https://doi.org/10.2118/109808-PA

65. Gu Y, Oliver DS (2007) An iterative ensemble Kalman filter for multiphase fluid flow data assimilation. SPE J 12(04):438–446. https://doi.org/10.2118/108438-PA

66. Dovera L, della Rossa E (2011) Multimodal ensemble Kalman filtering using Gaussian mixture models. Undefined 15(2):307–323. https://doi.org/10.1007/S10596-010-9205-3

67. Chen Y, Oliver DS, Zhang D (2009) Data assimilation for nonlinear problems by ensemble Kalman filter with reparameterization. J Petrol Sci Eng 66(1–2):1–14. https://doi.org/10.1016/J.PETROL.2008.12.002

68. Jahanbakhshi S, Pishvaie MR, Boozarjomehry RB (2018) Impact of initial ensembles on posterior distribution of ensemble-based assimilation methods. J Petrol Sci Eng 171:82–98. https://doi.org/10.1016/J.PETROL.2018.07.022

69. Emerick AA, Reynolds AC (2013) Investigation of the sampling performance of ensemble-based methods with a simple reservoir model. Comput Geosci 17(2):325–350. https://doi.org/10.1007/S10596-012-9333-Z

70. Lorentzen RJ, Naevdal G, Valles B, Berg A, Grimstad A-A (2005) Analysis of the ensemble Kalman filter for estimation of permeability and porosity in reservoir models. In: SPE Annual Technical Conference and Exhibition. Dallas, Texas, USA: OnePetro. https://doi.org/10.2118/96375-MS

71. Klema VC, Laub AJ (1980) The singular value decomposition: its computation and some applications. IEEE Trans Autom Control 25(2):164–176. https://doi.org/10.1109/TAC.1980.1102314

72. Chen Y, Oliver DS (2013) History matching of the norne full field model using an iterative ensemble smoother. In: 75th European association of geoscientists and engineers conference and exhibition 2013 incorporating SPE EUROPEC 2013: changing frontiers. European association of geoscientists and engineers, EAGE, pp 5730–5748. https://doi.org/10.2118/164902-MS

73. Evensen G, Raanes PN, Stordal AS, Hove J (2019) Efficient implementation of an iterative ensemble smoother for data assimilation and reservoir history matching. Frontiers Appl Math Stat 5:47. https://doi.org/10.3389/FAMS.2019.00047/BIBTEX

74. Emerick AA, Reynolds AC (2012) History matching time-lapse seismic data using the ensemble Kalman filter with multiple data assimilations. Comput Geosci 16(3):639–659. https://doi.org/10.1007/S10596-012-9275-5

75. Lee K, Jung S, Shin H, Choe J (2014) Uncertainty quantification of channelized reservoir using ensemble smoother with selective measurement data. Energy Explor Exploit 32(5):805–816

76. Liao Q, Alsamadony K, Lei G, Awotunde A, Patil S (2021) Reservoir history matching by ensemble smoother with principle component and sensitivity analysis for heterogeneous formations. J Petrol Sci Eng 198:108140. https://doi.org/10.1016/j.petrol.2020.108140

77. Reynolds AC, Zafari M, Li G (2006) Iterative forms of the ensemble Kalman filter. In: ECMOR 2006—10th European conference on the mathematics of oil recovery, cp-23-00025. https://doi.org/10.3997/2214-4609.201402496/CITE/REFWORKS

78. Goldberg DE (1989) Genetic algorithms in search, optimization and machine learning. Addison-Wesley Longman Publishing Co., Inc, Boston

79. Xavier CR, Dos Santos EP, Da Fonseca VV, Dos Santos RW (2013) Genetic algorithm for the history matching problem. Proc Comput Sci 18:946–955. https://doi.org/10.1016/J.PROCS.2013.05.260

80. Lange AG (2009) Assisted history matching for the characterization of fractured reservoirs. AAPG Bull 93(11):1609–1619. https://doi.org/10.1306/08040909050

81. Ballester PJ, Carter JN (2007) A parallel real-coded genetic algorithm for history matching and its application to a real petroleum reservoir. J Petrol Sci Eng 59(3–4):157–168. https://doi.org/10.1016/J.PETROL.2007.03.012

82. Chakra NCC, Saraf DN (2016) History matching of petroleum reservoirs employing adaptive genetic algorithm. J Petrol Explor Prod Technol 6(4):653–674. https://doi.org/10.1007/s13202-015-0216-4

83. Dermanaki Farahani Z, Ahmadi M, Sharifi M (2019) History matching and uncertainty quantification for velocity dependent relative permeability parameters in a gas condensate reservoir. Arabian J Geosci 12(15):1–19. https://doi.org/10.1007/S12517-019-4603-X/TABLES/9

84. Vadicharla G, Sharma P, Gupta SK, Saraf DN (2021) History matching of an oil reservoir using non-dominated sorting genetic algorithm-II coupled with sequential Gaussian simulation. In: 2021 International conference on innovative computing, intelligent communication and smart electrical systems (ICSES), pp 1–6. https://doi.org/10.1109/ICSES52305.2021.9633849

85. Sayyafzadeh M, Haghighi M, Carter JN (2012) Regularization in history matching using multi-objective genetic algorithm and bayesian framework. SPE Europec/EAGE Annual Conference SPE-154544-MS. https://doi.org/10.2118/154544-MS

86. Khamehchi E, Zolfagharroshan M, Mahdiani MR (2020) A robust method for estimating the two-phase flow rate of oil and gas using wellhead data. J Petrol Explor Prod Technol 10(6):2335–2347. https://doi.org/10.1007/s13202-020-00897-2

87. Maschio C, Nakajima L, Schiozer DJ (2008) Production strategy optimization using genetic algorithm and quality map. Europec/EAGE Conference and Exhibition 9–12. http://www.one petro.org/mslib/app/Preview.do?paperNumber=SPE-113483-MS&societyCode=SPE. https://doi.org/10.2118/113483-MS

88. Chakraborty A, Kar AK (2017) Swarm intelligence: a review of algorithms. Model Optim Sci Technol 10(March):475–494. https://doi.org/10.1007/978-3-319-50920-4_19

89. Kennedy J, Eberhart R (1995) Particle swarm optimization. In: Proceedings of ICNN'95 - international conference on neural networks. Perth, pp 1942–1948. https://doi.org/10.1109/ICNN.1995.488968

90. Yousefzadeh R, Sharifi M, Rafiei Y, Ahmadi M (2021) Scenario reduction of realizations using fast marching method in robust well placement optimization of injectors. Natl Resour Res 30:2753–2775. https://doi.org/10.1007/s11053-021-09833-5

91. Awotunde A (2019) An improved particle swarm optimization for history-matched reservoir parameters. EAGE Subsurface Intelligence Workshop 2019 2019(1):1–5. https://doi.org/10.3997/2214-4609.2019X6105/CITE/REFWORKS

92. Zhang L, Yang J, Sun Y, Zhang X, Tian W, Dong Z. A Novel Approach for Reservoir Automatic History Matching Based on the Hybrid of Particle Swarm Optimization and Gravitational Search Algorithm. In: Springer Series in Geomechanics and Geoengineering. Springer, Singapore; 2020. p. 3423–3439. https://doi.org/10.1007/978-981-16-0761-5_321

93. Mohamed L, Christie M, Demyanov V. History Matching and Uncertainty Quantification: Multiobjective Particle Swarm Optimisation Approach. 73rd European Association of Geoscientists and Engineers Conference and Exhibition 2011: Unconventional Resources and the Role of Technology. Incorporating SPE EUROPEC 2011. 2011;4:2927–2943. https://doi.org/10.2118/143067-MS

94. Lee S, Karl, Stephen D. Field Application Study on Automatic History Matching Using Particle Swarm Optimization. Society of Petroleum Engineers - SPE Reservoir Characterisation and Simulation Conference and Exhibition 2019, RCSC 2019. 2019 Sep 17. https://doi.org/10.2118/196678-MS

95. Christie M, Eydinov D, Demyanov V, Talbot J, Arnold D, Shelkov V (2013) Use of Multi-Objective Algorithms in History Matching of a Real Field. Society of Petroleum Engineers - SPE Reservoir Simulation Symposium 2013(1):57–67. https://doi.org/10.2118/163580-MS

96. Mohamed L, Christie M, Demyanov V, Robert E, Kachuma D (2010) Application of Particle Swarms for History Matching in the Brugge Reservoir. Proceedings - SPE Annual Technical Conference and Exhibition. 6:4477–4492. https://doi.org/10.2118/135264-MS

97. Razghandi M, Dehghan A, Yousefzadeh R (2021) Application of particle swarm optimization and genetic algorithm for optimization of a southern Iranian oilfield. Journal of Petroleum Exploration and Production. 11(4):1781–1796. https://doi.org/10.1007/s13202-021-01120-6

Chapter 4
Dimensionality Reduction Methods Used in History Matching

Abstract As discussed in Sect. 3.4, one of the challenges in history matching is the high dimensionality (large number of model parameters) of the reservoir model realizations that raises two challenges: 1- more data assimilation iterations are required to get a satisfactory match, 2- preserving the geologic realism becomes harder as there are an infinite number of solutions that can match the actual production data. These challenges are more prominent when dealing with spatially distributed properties such as the permeability distribution. Thereby, dimensionality reduction (also known as parametrization) methods are required to reduce the number of adjustable parameters while keeping the most salient ones. In the following, some of the dimensionality reduction methods used in the course of history matching are explained. These methods include the conventional methods, such as the pilot points, gradual deformation, principal component analysis, and higher-order singular value decomposition, and deep learning methods, including the autoencoders, variational autoencoders, and convolutional variational autoencoders. Also, a brief introduction to machine learning and deep learning is provided.

Keyword Dimensionality reduction · Parametrization · Machine learning · Deep learning · Autoencoder · Variational autoencoder · Pilot points · Gradual deformation · Principal component analysis

4.1 Conventional Dimensionality Reduction Methods

4.1.1 Pilot Points

In this method, instead of updating the value of all gridblocks/points, a reduced set of points is determined and updated by data assimilation. These points are known as the pilot points [1–4]. In fact, pilot points determine the locations where changes are made directly. For other points, the changes are made by lateral interpolation using kriging. Pilot points can be selected as the location of the wells. In the vertical direction, the changes are applied uniformly. It is possible to consider vertical correlation in the kriging as well. However, this asks for addition of additional pilot points in the vertical direction. Also, the correlation length in the vertical direction must be determined.

© The Author(s), under exclusive license to Springer Nature Switzerland AG 2023
R. Yousefzadeh et al., *Introduction to Geological Uncertainty Management in Reservoir Characterization and Optimization*, SpringerBriefs in Petroleum Geoscience & Engineering, https://doi.org/10.1007/978-3-031-28079-5_4

Pilot points with kriging was used by Kazemi and Stephen [1] to update the net/gross ratio, horizontal, and vertical permeability in history matching of Nelson reservoir model. They took benefit from the Neighborhood Algorithm as the optimizer and streamline simulation to assimilate data at each history matching iteration. Cartes and Marsily [5] also used the pilot points technique to identify aquifer transmissivities.

Although pilot points are applied to different history matching problems, it has two drawbacks including the optimal number of pilot points and their proper location, which have to be determined a priori [6].

4.1.2 Gradual Deformation

Gradual deformation is based on generating an ensemble of realizations and updating them smoothly at each step [7]. The update is performed based on a correlation between the realizations from one step to another. This is achieved by generating new realizations based on linear combinations of two previous realizations. Assuming Z_1 and Z_2 two independent prior realizations of a Gaussian random field, we can generate the posterior realization Z as follows:

$$Z = \cos(\rho\pi)Z_1 + \sin(\rho\pi)Z_2, \tag{4.1}$$

Note that we must normalize Z_1 and Z_2, with a zero mean, and not constrain them to the well data. Incorporation of the well data can be done by constraining the new realization Z. Finally, we have to superimpose the mean of the true distribution to the normalized Z. For example, the following transformation can be used to obtain a lognormal distribution:

$$Y = epx(\sigma Z + \mu), \tag{4.2}$$

were μ and σ denote the mean and the standard deviation of $\ln Y$, respectively. According to Eq. 4.1, only ρ is updated to generate a new realization which reduces the dimensionality of the problem, where ρ varies between -1 and $+1$. Equation 4.2 makes sure that the statistical moments of the new realization Z is honored for any ρ. The above parametrization allows a continuous control of generating new realizations. The initial realizations, Z_1 and Z_2, determine the domain of possible realizations. Varying ρ will lead to a continuous change in Z at each gridblock.

The flexibility of the methodology can be increased by incorporating more realizations to Eq. 4.1. In this manner, the new realization will be the linear combination of multiple initial realizations. Assuming $\{Z_1, Z_2, \ldots, Z_n\}$ be the set of n independent realizations of a Gaussian random field, the posterior realization can be generated using the following expression:

$$Z = \sum_{i=1}^{n} \beta_i Z_i, \tag{4.3}$$

where $\{\beta_1, \beta_2, \ldots, \beta_n\}$ represents the set of n real coefficients between -1 and 1.

In order to preserve the variogram of the model, the following condition must be satisfied:

$$\sum_{i=1}^{n} \beta_i^2 = 1. \tag{4.4}$$

It is common to honor this constraint where a variable transformation is used and the combination of realizations is done in spherical coordinates:

$$Z = \prod_{i=1}^{n-1} \cos(\rho_i \pi) Z_1 + \prod_{i=1}^{n-1} \sin(\rho_i \pi) \prod_{j=i+1}^{n-1} \sin(\rho_j \pi) Z_{i+1} + \ldots \tag{4.5}$$

In Eq. 4.5, the new realization Z is expressed as a function of $n - 1$ independent parameters $\{\rho_1, \rho_2, \ldots, \rho_n\}$. Thereby, the number of history matching parameters is reduced from the number of active grid cells in the models to the number or initial models minus one. If all parameters are zero, we get realization Z_1; otherwise, the result will be a linear combination of n independent realizations.

There are several combination schemes in gradual deformation to generate new realizations. One of the schemes is presented in [8] (see Fig. 3 in [8]). In this scheme, at each stage, each new realization is generated by implementing Eq. 4.1 on two previous realizations. Therefore, a single best-conditioned realization is obtained in the end. Thereby, to obtain an ensemble of conditioned models, one should run the gradual deformation several times. The optimal value of ρ is determined through an optimization process where the misfit function between the observed and predicted production data is minimized.

According to above explanations, the primary advantage of the gradual deformation is that it can update the initial (prior) models continuously, while reducing the dimensions of the problem. Also, it is able to preserve the geostatistical moments of the models during the deformation. This method can be regarded as a way of moving in the domain of possible realizations [7]. This method has been widely used in the course of history matching [7, 9–13]. In recent years, Heidari et al. [8] combined the gradual deformation with pressure derivative plots to history match the permeability distribution. They assumed that each region of the reservoir must be matched to a specific region of the derivative plot. Their main idea was to determine the pressure drop from the derivative plot and find the corresponding region of investigation (ROI) in the reservoir model using a simulator. Equation 4.6 gives the normalized pressure drop used in [8]:

$$\Delta p = 0.122 q \Delta p'. \tag{4.6}$$

Then, the derivative plot is divided into some regions (different times) and the corresponding pressure drop is calculated from Eq. 4.6. If the reservoir model is simulated, the pressure contour associated with each pressure drop determines the ROI corresponding to that pressure drop and time. Figure 12 in [8] shows the schematic of dividing the pressure derivative curve into 3 periods and the corresponding ROIs [8]. The permeability is updated using the gradual deformation where the deformation factor is optimized by minimizing the misfit function between the simulated and actual pressure derivative curves. However, this procedure is limited to 2D models with a Gaussian distribution of the permeability. Investigations can be done to generalize this method to 3D and non-Gaussian distributions.

4.1.3 Principle Component Analysis-based (PCA-based) Methods

Principal Component Analysis (PCA) [14] is a dimensionality reduction technique which assumes that a vector $v \in \mathbb{R}^n$ is constructed by a linear combination of several principal components. Consequently, by extracting the principal components and keeping the most important ones, we can reduce the dimension of a set of variables. What PCA does is reducing the correlation between the variables by producing new uncorrelated variables [6]. If $\mathbf{R} = \{\mathbf{r}_1, \ldots, \mathbf{r}_M\} \in \mathbb{R}^{N \times M}$ is a set of M realization vectors each of which with N elements, theses realizations first need to be zero-centered. We can define the set of centered realizations as $\mathbf{R} = \{\mathbf{r}_1 - \bar{\mathbf{R}}, \ldots, \mathbf{r}_M - \bar{\mathbf{R}}\}$ in which $\bar{\mathbf{R}} = \frac{1}{M} \sum_{i=1}^{M} r_i$. Therefore, the expectation and auto-covariance of $\mathbf{R}$ are given as follows:

$$E[\mathbf{R}] = 0 \tag{4.7}$$

$$E[\mathbf{R}\mathbf{R}^T] = \mathbf{C}_{RR}, \tag{4.8}$$

in which $\mathbf{C}_{RR}$ denotes the covariance matrix of geological parameters [6]. Actually, there is no analytical covariance matrix for $\mathbf{r}$, but it can be calculated using numerical methods:

$$\mathbf{C}_{RR} = E\left[(\mathbf{r} - E[\mathbf{r}])(\mathbf{r} - E[\mathbf{r}])^T\right] = \frac{1}{M-1}$$

$$\mathbf{R}\mathbf{R}^T = \frac{1}{M-1} \sum_{i=1}^{M} (\mathbf{r}_i - \bar{\mathbf{R}})(\mathbf{r}_i - \bar{\mathbf{R}})^T. \tag{4.9}$$

The goal of the PCA is to find a new set of parameters $\mathbf{y} = [y_1, \ldots, y_k]^T \in \mathbb{R}^k$ known as score variables, taking benefit from a linear transformation applied to the original parameters $\mathbf{r}$:

$$\mathbf{y} = \mathbf{P}^T (\mathbf{r} - \bar{\mathbf{R}}), \qquad (4.10)$$

where $\mathbf{P} = \begin{bmatrix} \boldsymbol{p}_1, \ldots, \boldsymbol{p}_k \end{bmatrix} \in \mathbb{R}^{N \times k}$ constrained in a way that:

$$\sqrt{\sum_{i=1}^{n} \mathbf{p}_{j,i}^2} = \|\mathbf{p}_j\|_2 = 1, \quad \text{for } j = 1, \ldots, k. \qquad (4.11)$$

in which $\|\mathbf{X}_2\| = \sqrt{x_1^2 + x_2^2 + \ldots + x_n^2}$ is the L2-norm of $\mathbf{X}$. If $\mathbf{r}$ is not zero-centered, an intercept vector is added to Eq. 4.9 [15]. Equations 4.12 and 4.13 show the expectation and covariance of $\mathbf{Y} = \begin{bmatrix} \boldsymbol{y}_1, \ldots, \boldsymbol{y}_m \end{bmatrix}$:

$$E[\mathbf{Y}] = 0, \qquad (4.12)$$

$$E\left[\mathbf{Y}\mathbf{Y}^T \right] = \boldsymbol{\Sigma}, \qquad (4.13)$$

where $\boldsymbol{\Sigma} \in \mathbb{R}^{k \times k}$ denotes the diagonal covariance matrix of $\mathbf{Y}$ which means the parameters in $\mathbf{Y}$ have the maximum variance with no correlation. The elements of $\mathbf{P} = \begin{bmatrix} \boldsymbol{p}_1, \ldots, \boldsymbol{p}_k \end{bmatrix} \in \mathbb{R}^{N \times k}$ are found by maximizing the variance as follows:

$$\lambda_i = \arg \max_{\mathbf{p}_i} \{ \mathbf{y}_i^T \mathbf{y}_i \} = \arg \max_{\mathbf{p}_i} \{ \mathbf{p}_i^T \mathbf{R}\mathbf{R}^T \mathbf{p}_i \}, \quad \text{for } i = 1, \ldots, k, \qquad (4.14)$$

subject to $\|\mathbf{p}_i\|_2^2 - 1 = 0$.

The constrained optimization problem in Eq. 4.14 can be converted to an unconstrained problem [16]:

$$\lambda_i = \arg \max_{\mathbf{p}_i} \{ E\left[\mathbf{p}_i^T \mathbf{R}\mathbf{R}^T \mathbf{p}_i \right] - \lambda_i \left(\mathbf{p}_i^T \mathbf{p}_i - 1 \right) \}. \qquad (4.15)$$

We can find the maximum of Eq. 4.15 by equating its gradient with respect to $\mathbf{p}_i$ to zero:

$$\mathbf{p}_i = \arg \frac{\partial}{\partial \mathbf{p}} \{ E\{ \mathbf{p}^T \mathbf{R}\mathbf{R}^T \mathbf{p} \} - \lambda_i \left(\mathbf{p}^T \mathbf{p} - 1 \right) \} = 0, \qquad (4.16)$$

Rearranging gives Eq. 4.17:

$$[\mathbf{C}_{RR} - \lambda_i \mathbf{I}] \mathbf{p}_i = 0. \qquad (4.17)$$

From Eq. 4.17 we can infer that the ith eigenvalue of $\mathbf{C}_{RR}$ is equivalent to the ith score variable. Also, the ith eigenvector of $\mathbf{C}_{RR}$ is equivalent to the ith vector of the transformation matrix. Sine $\mathbf{C}_{RR}$ is a positive definite matrix, we can conclude that the eigenvalues of $\mathbf{C}_{RR}$ are real and positive, and eigenvectors are reciprocally orthogonal. This means that there is a statistical independence between the score variables and their correlation is zero. The equivalence between the covariance matrix and a diagonal matrix of variances of score variables is proved in Eq. 4.18.

$$C_{YY} = \frac{1}{M-1}YY^T, \tag{4.18a}$$

If Eq. 4.10 is substituted into Eq. 4.18a for all realizations, it gives:

$$C_{YY} = \frac{1}{M-1}P^T R(P^T R)^T = \frac{1}{M-1}P^T RR^T P$$

$$= P^T \left(\frac{1}{M-1}RR^T\right)P = P^T C_{RR}P, \tag{4.18b}$$

Replacing the eigenvalue decomposition of C_{RR} (i.e., $C_{RR} = P\Sigma P^T$) with C_{RR} gives:

$$C_{YY} = P^T(P\Sigma P^T)P = (P^T P)\Sigma(P^T P), \tag{4.18c}$$

Because of the orthogonality of matrix P, Eq. 4.18c can be written as:

$$C_{YY} = \Sigma. \tag{4.18d}$$

Using the covariance (4.18d) above, we can generate a new realization of model parameters as follows:

$$r = \bar{R} + \frac{1}{\sqrt{N_R - 1}}U\Sigma z, \tag{4.19}$$

where z is an N-dimensional random Gaussian noise (i.e., $z \sim \mathcal{N}(0, I)$), and U is the matrix of the left singular vectors of $\bar{R}$.

It should be pointed out that the independency of score variables is true if the original parameters have a Gaussian distribution. Therefore, PCA is only applicable to problems where the distribution of the uncertain parameters follows a Gaussian distribution. Σ is a matrix of size $k \times k$. This indicates that a maximum of k dependent variables (score variables) are there (i.e., the maximum dimension of Y is k). Since some eigenvalues can be zeros, $k \leq M$. Also, each eigenvector having a zero eigenvalue is removed from P which means that the number of score variables is always equal or less than the number of realizations. In history matching, we can use these score variables as the matching variables (updated during data assimilation) to generate new ys. Then, y vectors can be used to generate new realizations using Eq. 4.10. Note that by eliminating eigenvectors with small eigenvalues (low variance), we can reduce the dimensionality of the adjustable parameters.

Esmaeili et al. [6] used a two-dimensional kernel-based PCA (2DKPCA) to reduce the dimension of the permeability realizations in the context of history matching. Since the standard PCA can only capture linear and Gaussian dependencies between the variables, Esmaeili et al. [6] suggested to use a kernel and transfer the data from the original space to a new space where they are linearly separable. Then, the standard PCA can be applied to reduce its dimensionality. Figure 2 in [6] shows the basic idea behind the kernel-PCA (KPCA). Also, they suggested to apply the PCA on rows and columns of the input matrices to avoid converting the 2D data into 1D vectors. Although introducing a kernel to PCA improved its results on non-Gaussian models, it needs to select a proper kernel. Also, the proposed methodology used a technique called the "pre-image" to back transfer the new realizations from the kernel space to the original space. This method is iterative and is not guaranteed to converge [6].

Emerick [17] employed three forms of PCAs including the standard PCA, kernel PCA (KPCA) and optimization-based PCA (OPCA) to parameterize facies realizations to be used for history matching by ES-MDA. OPCA converts the parameterization problem to an optimization problem. Equation 4.19 was rewritten by Vo and Durlofsky [18] in an optimization form as follows:

$$\mathbf{r} = \arg\min_{\hat{\mathbf{r}}} \left\| \bar{\mathbf{R}} + \frac{1}{\sqrt{N_R - 1}} \mathbf{U} \mathbf{\Sigma} \mathbf{z} - \hat{\mathbf{R}} \right\|_2^2. \tag{4.20}$$

If one wants to generate a binary model realization, a regularization term can be added to Eq. 4.20 as suggested in [18]. This regularization term converts the optimization problem to a constrained optimization problem:

$$\mathbf{r} = \arg\min_{\hat{\mathbf{r}}} \left\| \bar{\mathbf{R}} + \frac{1}{\sqrt{N_R - 1}} \mathbf{U} \mathbf{\Sigma} \mathbf{z} - \hat{\mathbf{r}} \right\|_2^2 + \gamma \hat{\mathbf{r}}^{\mathrm{T}} (1 - \hat{\mathbf{r}}),$$

subject to $\hat{r}_i \in [0, 1]$, for all i. \hfill (4.21)

in which $\gamma \in [0, 1]$ denotes the regularization coefficient. The term $\gamma \hat{\mathbf{r}}^{\mathrm{T}} (1 - \hat{\mathbf{r}})$ constrains each parameter of $\hat{\mathbf{r}}$ to either 0 or 1. Assuming that $\mathbf{r}_{\mathrm{pca}} = \bar{\mathbf{R}} + \frac{1}{\sqrt{N_R - 1}} \mathbf{U} \mathbf{\Sigma} \mathbf{z}$ is the solution of the PCA problem; therefore, the minimization problem can be rewritten as:

$$r_i = \arg\min_{\hat{r}_i} \left\{ (1 - \gamma)\hat{r}_i^2 - 2\left(r_{\mathrm{pca},i} - \frac{\gamma}{2}\right)\hat{r}_i \right\},$$

subject to $\hat{r}_i \in [0, 1]$, for all i. \hfill (4.22)

The solution to Eq. 4.22 is given by:

$$\hat{r}_i = 0, \text{ if } r_{\mathrm{pca},i} \leq \frac{\gamma}{2}$$

$$r_i = 1, \text{ if } r_{\mathrm{pca},i} \geq 1 - \frac{\gamma}{2}$$

$$\hat{r}_i = \frac{r_{\mathrm{pca},i} - \frac{\gamma}{2}}{1 - \gamma}, \text{ if } \frac{\gamma}{2} < r_{\mathrm{pca},i} < 1 - \frac{\gamma}{2}. \tag{4.23}$$

Emerick [17] used the PCA-based methods for parametrization of realizations of a 2D channelized model and concluded that the OPCA outperformed both the standard PCA and KPCA in terms of data match and similarity of the posterior realizations to the true case. Also, the results showed that the standard PCA did not improve the results of ES-MDA without any parametrization. Although OPCA improved the results of history matching, it was limited to two-facies models [17].

4.1.4 Higher-Order Singular Value Decomposition (HOSVD)

Higher-order singular value decomposition, also known as multi-linear singular value decomposition, is a form of singular value decomposition (SVD) for higher dimensional tensors [19]. This method is a tensor decomposition method that converts tensors to matrix elements. HOSVD of a 3D tensor $\mathbf{X}$ entails the singular value decomposition of three matrices, known as the matrix modes [20]. The workflow of the HOSVD for a 3D tensor, $\mathbf{X}$, can be summarized in five steps as follows [19, 20]:

1. Unfold the tensor $\mathbf{X} \in \mathbb{R}^{m \times n \times p}$, where three matrices (1-mode $\mathrm{U}^{(1)}$, 2-mode $\mathrm{U}^{(2)}$, 3-mode $\mathrm{U}^{(3)}$) are built:

$$U_1 \in \mathbb{R}^{I_1 \times (I_2 I_3)}, U_1 \in \mathbb{R}^{I_2 \times (I_1 I_3)}, U_1 \in \mathbb{R}^{(I_1 I_2) \times I_3}. \tag{4.24}$$

2. Apply SVD on the three matrices obtained from step 1.
3. From each mode, keep k-most important singular vectors corresponding to the largest singular values.
4. Reduce the dimensionality of the tensor by creating the core tensor:

$$\mathbf{S} = \mathbf{X} \times_1 \mathrm{U}^{(1)^{\mathrm{T}}} \times_2 \mathrm{U}^{(2)^{\mathrm{T}}} \times_3 \mathrm{U}^{(3)^{\mathrm{T}}}. \tag{4.25}$$

5. Construct the approximation tensor:

$$\hat{X} = \mathbf{S} \times_1 \mathrm{U}^{(1)} \times_2 \mathrm{U}^{(2)} \times_3 \mathrm{U}^{(3)}. \tag{4.26}$$

Vaseghi et al. [20] employed the HOSVD to reduce the dimensionality of realizations to use in the K-means clustering. Afra and Gildin also used the HOSVD to compress permeability maps and compared the results with the results of the standard PCA. They found that the HOSVD performs better than the standard PCA by comparing the reconstructions of the two methods.

There are other parametrization methods, such as Discrete Cosine Transform (DCT) and Discrete Wavelet Transform (DWT), which use some coefficients to parametrize the input. The details of these methods are beyond the scope of this book. Interested readers can refer to [21–23] for more details on these methods.

4.2 Machine Learning-based Methods

4.2.1 Introduction to Machine Learning and its Applications

Machine learning (ML) is a branch of artificial intelligence that gives the computers the ability to learn from data without being explicitly programmed. The data used for building the machine learning models are called training data/samples. A formal definition to machine learning is given by Mitchell [24]: "A computer program is said to learn from experience E with respect to some task T and some performance measure P, if its performance on T, as measured by P, improves with experience E". There are four main categories of learning in ML, including the supervised, unsupervised, semi-supervised, and reinforcement learning. In supervised learning, the training data are provided as pairs of (inputs, outputs/labels), and the model learns the relationship between the inputs and outputs, whereas no labels are provided in unsupervised learning, and the goal is to extract the hidden structures in the inputs. Semi-supervised learning refers to problems where some of the data have been provided with labels, and some of them do not have labels. In Reinforcement learning, an agent interacts with an environment and receives reward from the environment. The goal of the agent is to learn a policy that maximizes the received rewards [25].

Machine learning has had successful applications in a wide variety of problems, such as speech recognition [26], computer vision [27], natural language processing [28], and agriculture [29] among others. Application of ML techniques is not exceptional for petroleum engineering and numerous classification, regression, and clustering tasks have been solved using the ML algorithms. For example, different ML algorithms were used in [30] to predict the scale formation in oil/gas wells, predict frictional loss in inclined annuli [31], estimate the light oil viscosity [32], predict the amount of wax precipitation in reservoirs [33], predict formation damage [34], classify rock types [35], core-log interpretation for deriving permeability and reservoir rock typing [36], etc.

4.2.2 Introduction to Deep Learning and its Applications

Deep learning (DL) is a branch of machine learning that is based on deep Artificial Neural Networks (ANNs) with several hidden layers [25]. DL algorithms mitigate the need for manual feature extraction from the input data by automating this process. Therefore, DL algorithms can perform better than other ML methods since DL algorithms can better identify and extract hidden dependencies in the input. Convolutional Neural Networks (CNNs), Recurrent Neural Networks (RNNs), Long Short-Term Memory (LSTM), Gated Recurrent Units (GRUs), Autoencoders (AEs), Variational Autoencoders (VAEs), and Generative Adversarial Networks (GANs) are some of the deep learning architectures. Each DL method has had successful applications in reservoir engineering, for example, CNNs have been used for well placement optimization [37, 38] and history matching [39], LSTM and RNN have been used for oil price [40, 41] and production [42, 43] prediction. AEs and VAEs are unsupervised learning algorithms that are used to extract the most salient features from the input. Therefore, the extracted features can be used instead of the original input with the aim of dimensionality reduction. GANs also can be used for dimensionality reduction by learning the data generating distribution of the training data. Application of the DL for dimensionality reduction of geological realizations is described in the following.

4.3   Deep Learning Methods used for Dimensionality Reduction of Geological Realizations

Deep-learning-based methods are used to capture the non-linear, complex, and non-Gaussian structure of the geological realizations. In this regard, the procedure for training a deep model for parameterization of the realizations follows an unsupervised learning method. This is because, the parameterization step is performed prior to starting any history matching process. As a consequence, Autoencoders, and their extensions, have gained more traction in recent years for parameterization and dimensionality reduction of high-dimensional reservoir models.

4.3.1   Autoencoders

Autoencoder is an end-to-end neural network that consists of two parts: encoder and decoder. The encoder receives the vector of inputs, multiplies it by some weights, and maps it into a low-dimensional feature space, also known as the latent space. Then, the decoder gets the features from the latent space and reconstructs the input in the output layer. The dimension of the input and output is the same in an AE [25, 44, 45]. Figure 4.1 demonstrates the architecture of an AE. As Fig. 4.1 shows, an AE has three layers: input, hidden, and output layers. Green and orange circles represent the input and output features, and the blue rectangles represent the latent space. The latent space usually has a lower dimension in comparison to the input. This characteristic of the AE leads to learning the most salient features of the input. An AE is trained in an unsupervised manner where the goal is to minimize the reconstruction error between the reconstructed input and the original input. Of course, a well-trained AE is an AE that results in a low reconstruction error. Equation 4.27 gives the objective function of an AE.

$$\mathcal{L}_{RE}(m) = \frac{1}{N_m} \sum_{i=1}^{N_m} \sum_{j=1}^{N_f} \left(m_{ij} - \hat{m}_{ij}\right)^2, \tag{4.27}$$

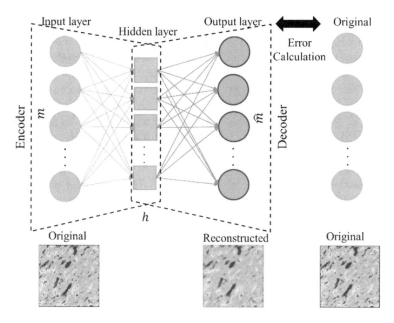

Fig. 4.1 Architecture of an Autoencoder

where m is the original input, $\hat{m}$ is the reconstructed input (output), N_m is the number of samples, and N_f is the number of features (i.e., dimension of the input). Equation 4.27 can be regularized to avoid overfitting. Therefore, Eq. 4.27 can be written as Eq. 4.28 if one adds the weight decay and sparsity regularization terms.

$$\mathcal{L}_{RE}(m) = \frac{1}{N_m} \sum_{i=1}^{N_m} \sum_{j=1}^{N_f} (m_{ij} - \hat{m}_{ij})^2 + \lambda \Omega_w + \beta \Omega_s, \qquad (4.28)$$

where Ω_w is the sum of squared weights (except for the bias), λ is the weight decay regularization coefficient, and β is the sparsity regularization coefficient. Ω_w and Ω_s can be determined as follows:

$$\Omega_w = \frac{1}{2} \sum_{i=1}^{N_m} \sum_{j=1}^{N_f} (w_{ij})^2, \qquad (4.29)$$

$$\Omega_s = \sum_{k=1}^{N_{node}} \rho \log\left(\frac{\rho}{\hat{\rho}_k}\right) + (1 - \rho)\left(\frac{1 - \rho}{1 - \hat{\rho}_k}\right), \qquad (4.30)$$

$$\hat{\rho}_k = \frac{1}{N_m} \sum_{i=}^{N_m} h_k(m_i), \qquad (4.31)$$

where w_{ij} is the weight of the jth feature of the ith sample, N_{node} is the number of neurons in a layer, ρ is a desired average value for the output of a layer, $\hat{\rho}_k$ is the average value of the kth neuron in the hidden layer, and $h_k(m_i)$ is the value of the kth neuron in the hidden layer. We can add more hidden layers to an AE to be able to capture more non-linear structures, which is known as the Stacked Autoencoder (SAE).

Kim et al. [46] used two extensions of the AE, including the SAE and Denoising Serial AE (SDAE), combined with ES-MDA for history matching of facies realizations. The role of the SDAE was to remove two types of noises, including the Gaussian noise and the salt-and-pepper noise from the output of the ES-MDA. Also, SAE was used to parametrize the facies realizations. The proposed method improved upon the results in terms of pushing the histogram of the posterior models toward the Bernoulli distribution (Fig. 4 in [46]).

Zhang et al. [47] also used a Deep Sparse Autoencoder (DSAE) for parameterization of fractures in naturally fractured reservoirs. They represented each fracture using three parameters (length, orientation, and midpoint coordinates) to generate a network of fractures. Since the number of parameters increases as the number of fractures gets higher, they employed the proposed DSAE for dimensionality reduction of the problem. They used a greedy layer-wise pretraining followed by a fine-tuning to train the network. For more details on the greedy layer-wise pretraining and fine tuning please refer to [25]. Zhang et al. [47] compared the results with the stanadard PCA and no-parameterization cases and found that when the number of fractures is low, all three methods give similar results; however, for complex models with a larger number of fractures, DSAE gives the best results. Also, in complex reservoirs, PCA and no-parameterization methods did not improve the history matching results.

4.3.2 Variational Autoencoders (VAEs)

In standard AEs we do not have any prior information about the distribution of the latent space. Consequently, we cannot draw a sample out of this distribution. Thereby, standard AEs are not able to generate new samples. VAEs [25, 48] have an additional layer which is responsible for generating new samples. Contrary to the standard AEs that encode the input to an unknown latent space, z, in VAEs a distribution is presumed for the latent space; thereby, we can sample from this distribution. Accordingly, VAEs have an additional term in their objective function that quantifies the mismatch between the distribution of the encoded latent space, $p(z|m)$, and the presumed distribution $p(z)$. This distribution can be a simple normal-standard distribution $\mathcal{N}(0, I)$. This yields more continuous latent space in comparison to the latent space of the standard AEs. In order to calculate the difference between two distributions, the Kullback–Leibler divergence is used [25]. Therefore, the loss function of the VAE is given as:

$$\mathcal{L}(m) = \mathcal{L}_{RE}(m) + \mathcal{D}_{KL}(p(z|m)\|p(z)), \qquad (4.32)$$

in which, $\mathcal{L}(\boldsymbol{m})$ is the overall loss function, $\mathcal{L}_{RE}(\boldsymbol{m})$ is the reconstruction loss (Eq. 4.27), $\mathcal{D}_{KL}(p(\boldsymbol{z}|\boldsymbol{m})||p(\boldsymbol{z}))$ is the Kullback–Leibler divergence between two distributions $p(\boldsymbol{z}|\boldsymbol{m})$ and $p(\boldsymbol{z})$. $\mathcal{D}_{KL}(p(\boldsymbol{z}|\boldsymbol{m})||p(\boldsymbol{z}))$ acts like a regularization term which is imposed on the latent space. Assuming $p(\boldsymbol{z}|\boldsymbol{m}) = \mathcal{N}\left(\left[\mu_1, \ldots, \mu_{N_z}\right]^T, diag\left[\sigma_1^2, \ldots, \sigma_{N_z}^2\right]^T\right)$ and $p(\boldsymbol{z}) = \mathcal{N}(0, I)$, the Kullback–Leibler divergence is defined as follows:

$$\mathcal{D}_{KL}(p(\boldsymbol{z}|\boldsymbol{m})||p(\boldsymbol{z})) = \frac{1}{2}\sum_{i=1}^{N_z}(\mu_i^2 + \sigma_i^2 - \ln(\sigma_i^2) - 1), \qquad (4.33)$$

where μ and σ are the vectors of mean and standard deviation of the latent space, and N_z is the number of latent features, which is much fewer than the original dimension of the input. When training the network, instead of generating z, μ and $\ln(\sigma^2)$ are predicted by the encoder. Using these values and a random noise $\hat{z} \sim \mathcal{N}(0, I)$, the vector of latent features is generated:

$$z = \mu + \sigma \odot \hat{z}. \qquad (4.34)$$

The vector of latent features is then given to the decoder that reconstructs the input in the output. After the network is trained, its decoder part can be used to reconstruct new realizations from low-dimensional space to the original space. Therefore, we can only update the low-dimensional features in the history matching process instead of high-dimensional realizations. Figure 4.2 shows the schematic of a deep VAE.

Using VAEs for dimensionality reduction is a novel approach in history matching of petroleum reservoirs. Recently, Canchumuni et al. [49] used different VAE-based

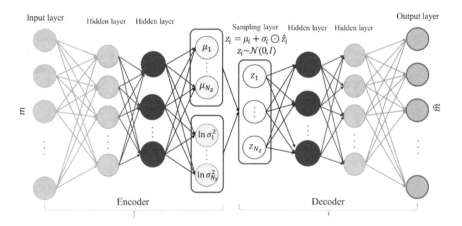

Fig. 4.2 Architecture of a deep Variational Autoencoder

deep networks for parameterization of 2D channelized facies models. The parameterized realizations were used in the context of history matching by the ES-MDA method. They compared the data match and the quality of generated realizations with the reference case for the VAE-based and other methods (e.g., GANs [25]) and concluded that the VAE-based methods had the best results with a low computational cost. The overall workflow of the history matching in combination with the dimensionality reduction is demonstrated in Fig. 7 of [49]. In this figure, the deep network encodes the prior realizations to the latent space only once, and in the rest of the history matching process, the decoder part reconstructs the low-dimensional realizations generated by the ES-MDA. The reconstructed realizations are given to a reservoir simulator to assimilate new production data.

4.3.3 Convolutional Variational Autoencoder (C-VAE)

Although the VAEs have shown good performance in parameterization of 2D facies models, they receive the input as 1D vectors that results in the loss of higher order statistics related to 2D and 3D dependencies. Therefore, instead of using fully connected layers in VAEs, convolutional layers can be used that can process image and volume data. The architecture of the C-VAE is similar to that of VAEs, but instead of fully connected layers, we use convolutional layers in the encoder and decoder parts of the network. However, the sampling layer and mean and standard deviation layers are the same as were in VAEs. Figure 3 in [50] shows the schematic of a C-VAE.

Canchumuni et al. [50] used a C-VAE for parametrization of channelized facies models to be used in history matching by the ES-MDA. To parameterize 3D models, they used three-dimensional convolutions. They compared the results with the OPCA and a Deep Belief Network [25] and found that the C-VAE outperformed the other methods in terms of the data match and plausibility of the history matched facies models. However, the results were not very satisfying on the prediction data (after history matching). This can be because of the inability of three-dimensional convolution in parameterizing 3D models.

References

1. Kazemi A, Stephen KD (2013) Mise à jour optimale des paramètres dans un processus de calage d'historique en s'aidant des lignes de courants. Oil Gas Sci Technol 68(3):577–594. https://doi.org/10.2516/ogst/2012071
2. De Marsily G, Lavedan G, Boucher M, Fasamino G (1984) Interpretation of interference tests in a well field using geostatistical techniques to fit the permeability distribution in a reservoir model. Geostatistics for natural resources characterization. NATO Adv Study Inst 831–849
3. LaVenue AM, Pickens JF (1992) Application of a coupled adjoint sensitivity and kriging approach to calibrate a groundwater flow model. Water Resour Res 28(6):1543–1569. https://doi.org/10.1029/92WR00208

4. RamaRao BS, LaVenue AM, De Marsily G, Marietta MG (1995) Pilot point methodology for automated calibration of an ensemble of conditionally simulated transmissivity fields: 1. Theory and computational experiments. Water Resour Res 31(3):475–493. https://doi.org/10.1029/94WR02258

5. Certes C, de Marsily G (1991) Application of the pilot point method to the identification of aquifer transmissivities. Adv Water Resour 14(5):284–300. https://doi.org/10.1016/0309-1708(91)90040-U

6. Esmaeili M, Ahmadi M, Kazemi A (2020) Kernel-based two-dimensional principal component analysis applied for parameterization in history matching. J Petrol Sci Eng 2020(191):107134. https://doi.org/10.1016/j.petrol.2020.107134

7. Roggero F, Hu LY (1998) Gradual deformation of continuous geostatistical models for history matching. In: SPE annual technical conference and exhtition 221–236. https://doi.org/10.2523/49004-ms

8. Heidari Sureshjani M, Ahmadi M, Fahimpour J (2020) Estimating reservoir permeability distribution from analysis of pressure/rate transient data: a regional approach. J Petrol Sci Eng 191:107172. https://doi.org/10.1016/J.PETROL.2020.107172

9. Gautier Y, Nœtinger B, Roggero F (2004) History matching using a streamline-based approach and gradual deformation. SPE J 9(01):88–101. https://doi.org/10.2118/87821-PA

10. Busby D, Feraille M, Gervais V (2009) Uncertainty reduction by production data assimilation combining gradual deformation with adaptive response surface methodology. 71st European Association of Geoscientists and Engineers Conference and Exhibition 2009: Balancing Global Resources. Incorporating SPE EUROPEC 2009. 6:3470–3481. https://doi.org/10.2118/121274-MS

11. Le Gallo Y, Le Ravalec-Dupin M (2000) History matching geostatistical reservoir models with gradual deformation method. In: SPE Annual Technical Conference and Exhibition. OnePetro. https://doi.org/10.2118/62922-MS

12. Neau A, Thore P, de Voogd B (2008) Combining the gradual deformation method with seismic forward modeling to constrain reservoir models. SEG Tech Program Expanded Abstracts 27(1):1910–1914. https://doi.org/10.1190/1.3059396

13. Le Ravalec-Dupin M, Nœtinger B (2002) Optimization with the gradual deformation method. Math Geol 34(2):125–142. https://doi.org/10.1023/A:1014408117518

14. Maćkiewicz A, Ratajczak W (1993) Principal components analysis (PCA). Comput Geosci 19(3):303–342. https://doi.org/10.1016/0098-3004(93)90090-R

15. Vidal R, Ma Y, Sastry SS (2016) Generalized principal component analysis. New York, NY: Springer New York. (Interdisciplinary Applied Mathematics). https://doi.org/10.1007/978-0-387-87811-9

16. Anderson TW (1962) An introduction to multivariate statistical analysis. Wiley

17. Emerick AA (2017) Investigation on principal component analysis parameterizations for history matching channelized facies models with ensemble-based data assimilation. Math Geosci 49(1):85–120. https://doi.org/10.1007/s11004-016-9659-5

18. Vo HX, Durlofsky LJ (2014) A new differentiable parameterization based on principal component analysis for the low-dimensional representation of complex geological models. Math Geosci 46(7):775–813. https://doi.org/10.1007/S11004-014-9541-2

19. Afra S, Gildin E (2013) Permeability parametrization using higher order singular value decomposition (HOSVD). In: Proceedings—2013 12th international conference on machine learning and applications, ICMLA 2013. 2:188–193. https://doi.org/10.1109/ICMLA.2013.121

20. Vaseghi F, Ahmadi M, Sharifi M, Vanhoucke M (2021) Generalized multi-scale stochastic reservoir opportunity index for enhanced well placement optimization under uncertainty in green and brown fields. Oil Gas Sci Technol 76(41). https://doi.org/10.2516/ogst/2021014

21. Kim S, Lee C, Lee K, Choe J (2015) Aquifer characterization of gas reservoirs using Ensemble Kalman filter and covariance localization. J Petrol Sci Eng 2016(146):446–456. https://doi.org/10.1016/j.petrol.2016.05.043

22. Zhao Y, Forouzanfar F, Reynolds AC (2017) History matching of multi-facies channelized reservoirs using ES-MDA with common basis DCT. Comput Geosci 21(5–6):1343–1364. https://doi.org/10.1007/s10596-016-9604-1

23. Luo X, Bhakta T, Jakobsen M, Nædal G (2016) An ensemble 4D seismic history matching framework with sparse representation based on wavelet multiresolution analysis. Soc Petrol Eng—SPE Bergen One Day Seminar. https://doi.org/10.2118/180025-ms

24. Mitchell TM (2010) Chapter 1 generative and discriminative classifiers: Naive Bayes and logistic regression learning classifiers based on Bayes rule. Mach Learn 1(Pt 1–2):1–17. https://doi.org/10.1093/bioinformatics/btq112

25. Goodfellow I, Bengio Y, Courville A (2016) Deep learning. The MIT Press

26. Vashisht V, Pandey AK, Yadav SP (2021) Speech recognition using machine learning. IEIE Trans Smart Process Comput 10(3):233–239. https://doi.org/10.5573/IEIESPC.2021.10.3.233

27. Khan AA, Laghari AA, Awan SA (2021) Machine learning in computer vision: a review. EAI Endorsed Tran Scalable Inf Syst 8(32):1–11. https://doi.org/10.4108/eai.21-4-2021.169418

28. Khurana D, Koli A, Khatter K, Singh S (2022) Natural language processing: state of the art, current trends and challenges. Multimedia Tools Appl. https://doi.org/10.1007/s11042-022-13428-4

29. Meshram V, Patil K, Meshram V, Hanchate D, Ramkteke SD (2021) Machine learning in agriculture domain: a state-of-art survey. Artific Intell Life Sci 1(September):100010. https://doi.org/10.1016/j.ailsci.2021.100010

30. Yousefzadeh R, Bemani A, Kazemi A, Ahmadi M (2022) An insight into the prediction of scale precipitation in harsh conditions using different machine learning algorithms. SPE Prod Oper 17:1–19. https://doi.org/10.2118/212846-PA

31. Bemani A, Kazemi A, Ahmadi M, Yousefzadeh R, Moraveji MK (2022) Rigorous modeling of frictional pressure loss in inclined annuli using artificial intelligence methods. J Petrol Sci Eng 211:110203. https://doi.org/10.1016/j.petrol.2022.110203

32. Lashkenari MS, Taghinezhad M, Mehdizadeh B (2013) Viscosity prediction in selected Iranian light oil reservoirs: artificial neural networks versus empirical correlations. Pet Sci 10:126–133

33. Khalighi J, Cheremisin A (2022) Wax precipitation prediction using a novel intelligent method: modeling and data analysis. Petrol Sci Technol 1–23. https://doi.org/10.1080/10916466.2022.2143800

34. Ahmadi M, Chen Z (2020) Machine learning-based models for predicting permeability impairment due to scale deposition. J Petrol Explor Prod Technol 10(7):2873–2884. https://doi.org/10.1007/s13202-020-00941-1

35. Mohammadian E, Kheirollahi M, Liu B, Ostadhassan M, Sabet M (2022) A case study of petrophysical rock typing and permeability prediction using machine learning in a heterogenous carbonate reservoir in Iran. Sci Rep 12(1):1–15. https://doi.org/10.1038/s41598-022-08575-5

36. Saputelli L, Celma R, Boyd D, Shebl H, Gomes J, Bahrini F, Escorcia A, Pandey Y (2019) Deriving permeability and reservoir rock typing supported with self-organized maps SOM and artificial neural networks ANN—optimal workflow for enabling core-log integration. SPE Reservoir Characterisation and Simulation Conference and Exhibition. https://doi.org/10.2118/196704-MS

37. Kwon S, Park G, Jang Y, Cho J, Chu M Gon, Min B (2021) Determination of oil well placement using convolutional neural network coupled with robust optimization under geological uncertainty. J Petrol Sci Eng 201(June 2020). https://doi.org/10.1016/j.petrol.2020.108118

38. Chu M Gon, Min B, Kwon S, Park G, Kim S, Huy NX (2020) Determination of an infill well placement using a data-driven multi-modal convolutional neural network. J Petrol Sci Eng 195(May):106805. https://doi.org/10.1016/j.petrol.2019.106805

39. Mohd Razak S, Jafarpour B (2020) Convolutional neural networks (CNN) for feature-based model calibration under uncertain geologic scenarios. Comput Geosci 24(4):1625–1649. https://doi.org/10.1007/s10596-020-09971-4

40. Yao T, Wang Z (2021) Crude oil price prediction based on LSTM network and GM (1,1) model. Grey Syst: Theory Appl 11(1):80–94. https://doi.org/10.1108/GS-03-2020-0031

41. Aziz N, Abdullah MHA, Zaidi AN (2020) Predictive analytics for crude oil price using RNN-LSTM neural network. In: 2020 International conference on computational intelligence (ICCI), 173–178. https://doi.org/10.1109/ICCI51257.2020.9247665

42. Sagheer A, Kotb M (2019) Time series forecasting of petroleum production using deep LSTM recurrent networks. Neurocomputing 323:203–213. https://doi.org/10.1016/j.neucom.2018.09.082

43. Temizel C, Canbaz CH, Saracoglu O, Putra D, Baser A, Erfando T, Krishna S, Saputelli L (2020) Production Forecasting in Shale Reservoirs Using LSTM Method in Deep Learning. https://doi.org/10.15530/urtec-2020-2878

44. Baldi P (2012) Autoencoders, unsupervised learning, and deep architectures. In: Proceedings of machine learning research, vol 27. JMLR Workshop and Conference Proceedings, pp 37–49.

45. Li W, Fu H, Yu L, Gong P, Feng D, Li C, Clinton N (2016) Stacked Autoencoder-based deep learning for remote-sensing image classification: a case study of African land-cover mapping. Int J Remote Sens 37(23):5632–5646. https://doi.org/10.1080/01431161.2016.1246775

46. Kim S, Min B, Kwon S, Chu MG (2019) History matching of a channelized reservoir using a serial denoising autoencoder integrated with ES-MDA. Geofluids. 2019. https://doi.org/10.1155/2019/3280961

47. Zhang K, Zhang J, Ma X, Yao C, Zhang L, Yang Y, Wang J, Yao J, Zhao H (2021) History matching of naturally fractured reservoirs using a deep sparse autoencoder. SPE J 26(04):1700–1721. https://doi.org/10.2118/205340-pa

48. Pu Y, Gan Z, Henao R, Yuan X, Li C, Stevens A, Carin L (2016) Variational autoencoder for deep learning of images, labels and captions. In: Advances in Neural Information Processing Systems, vol 29

49. Canchumuni SWA, Castro JDB, Potratz J, Emerick AA, Pacheco MAC (2020) Recent developments combining ensemble smoother and deep generative networks for facies history matching. https://doi.org/10.1007/s10596-020-10015-0

50. Canchumuni SWA, Emerick AA, Pacheco MAC (2019) Towards a robust parameterization for conditioning facies models using deep variational autoencoders and ensemble smoother. Comput Geosci 128(January):87–102. https://doi.org/10.1016/j.cageo.2019.04.006

Chapter 5
Field Development Optimization Under Geological Uncertainty

Abstract Decision making about field development plans has to consider the inherent uncertainties of sub-surface hydrocarbon reservoirs; therefore, the decisions would be stable under different geological scenarios. As described in Sect. 1.5.1, the aim of this kind of uncertainty management is to propagate the uncertainty from inputs to the outputs. Therefore, instead of a single deterministic output, the output will be probabilistic from which some statistical measures, such as the expected value, standard deviation, etc. can be calculated to account for the uncertainty. Therefore, the final decision regarding the field development plan can be taken according to the statistical measures. This kind of uncertainty management is also known as the robust field development optimization as explained in the following. In addition, different risk measures, different approaches to selecting an ensemble of representative geological realizations to be used in robust optimization, decreasing the computational cost of the optimization under geological uncertainty by constrained optimization, and challenges related to these activities are described in this chapter.

Keywords Robust optimization · Risk attitude · Representative realizations · Constrained optimization · Computational cost · Clustering · Geological uncertainty

5.1 Fundamentals of Robust Field Development Optimization

Robust field development optimization under geological uncertainty refers to making optimum decisions to develop an oil/gas field with the consideration of geological uncertainties. In this approach, geological uncertainty is captured using a series of geological realizations. Robust field development optimization can be formulated as follows:

For a set of solutions, $\mathbb{X}$, we seek for a solution $\mathbf{x}^*$ that maximizes:

$$J = E_M[f(\mathbf{x}, m)], \tag{5.1}$$

© The Author(s), under exclusive license to Springer Nature Switzerland AG 2023
R. Yousefzadeh et al., *Introduction to Geological Uncertainty Management in Reservoir Characterization and Optimization*, SpringerBriefs in Petroleum Geoscience & Engineering, https://doi.org/10.1007/978-3-031-28079-5_5

where $f(\mathbf{x}, m)$ is the value of the objective function on realization m with decision variables $\mathbf{x}$. To account for the geological uncertainty, the expected value of $f(\mathbf{x}, m)$ over M realizations has to be optimized. As a result, the final objective function of the problem would be $E_M[f(\mathbf{x}, m)]$. In this approach, each set of decision variables is applied on all available realizations to compute the expected value of $f(\mathbf{x}, m)$ over M realizations, which is defined as:

$$E_M[f(\mathbf{x}, m)] = \sum_{i=1}^{M} p(m_i) f(\mathbf{x}, m_i), \tag{5.2}$$

where $p(m_i)$ is the probability of the geological realization m_i. These probabilities can be obtained during sampling from the prior probability distribution of the uncertain parameters using sampling methods, such as the Monte Carlo simulation or MCMC methods. If the realizations are equiprobable, the expectation will be a simple arithmetic average. Robust field development optimization entails several challenges, which are presented in the following.

5.2 Challenges in Robust Field Development Optimization

As discussed above, in robust field development optimization (FDO), we need to calculate the objective function over multiple geological realizations and compute its expected value. This practice brings about some challenges, including the high number of reservoir simulations, selection of a proper risk attitude, and selection of a proper representative subset of geological realizations. These challenges are explained in detail in the following subsections.

5.2.1 High Number of Reservoir Simulations

One of the biggest challenges in robust field development optimization is the high number of reservoir simulations required to calculate the objective function. Since we need to evaluate the objective function over a large number of geological realizations, the overall computational cost and time will increase proportional to the number of realizations. As in almost all field development optimization problems that it is needed to conduct reservoir simulations, if the reservoir model has a large dimension, this increased computational cost will be more of concern [1].

5.2.2 Risk Attitudes

Our attitude toward risk is another challenge is robust FDO that greatly influences the ultimate results. The risk attitude determines the shape of the overall objective function. Risk attitude is defined as the way we are going to treat the existing risks in the context of robust FDO. Since we do not have access to the true reservoir model, we have to make a decision that does not change drastically if the reservoir model turns out to be different from the realizations involved in the development process. The risk attitude determines how to manage this risk. If we make our decision based on the expected value of the objective function, in fact, we take no risk management measures and only rely on the average performance of the approach. Several risk attitudes can be used to manage the risk associated with the task of robust FDO including the worst-case, best-case, expected profit, coefficient of variation, mean-standard deviation, semi-variance, mean-semi-variance, Value at Risk (VaR), Conditional Value at Risk (CVaR), minimax regret function, etc. These risk attitudes are discussed below based on the studies from [2].

Worst-case
In this risk attitude, only the profit of the worst case is considered as the objective function. As a result, this risk measure is very conservative specially when the truth is very different from the worst-case.

$$\mathcal{R}(f(\mathbf{x}, m)) = \min_{m} f(\mathbf{x}, m). \tag{5.3}$$

Best-case
The best-case risk measure considers only the best profit and ignores other realizations. This risk measure is very optimistic if the best-case realization is very different from the truth.

$$\mathcal{R}(f(\mathbf{x}, m)) = \max_{m} f(\mathbf{x}, m). \tag{5.4}$$

Expected profit
This risk measure only considers the expectation over the realizations. It was first used in the petroleum industry by Van Essen et al. [2] to optimize a water-flooding project under geological uncertainty, which is the same as Eq. 5.1.

$$E_M[f(\mathbf{x}, m)] = \sum_{i=1}^{M} p(m_i) f(\mathbf{x}, m_i). \tag{5.5}$$

Coefficient of Variation
This measure is the ratio of the expected profit to the standard deviation of the objective function over the realizations. If the overall objective function is maximized,

the expectation is increased and the standard deviation is decreased. This helps to come up with solutions that does not vary drastically with the variation of the reservoir model.

$$CV[f(\mathbf{x}, m)] = E_M[f(\mathbf{x}, m)]/\sigma(f(\mathbf{x}, m)), \tag{5.6}$$

where σ is the standard deviation of the objective function over the realizations.

Mean-Standard Deviation
Another way to decrease the standard deviation is to use the mean-standard deviation risk measure. In this approach, the standard deviation is subtracted from the expected value. Therefore, we can maximize the expectation and minimize the standard deviation at the same time.

$$J = E_M[f(\mathbf{x}, m)] - \sigma(f(\mathbf{x}, m)). \tag{5.7}$$

Most of the time, this risk measure comes with a risk factor which varies between -1 and 1. Therefore, the general formulation will be:

$$J = E_M[f(\mathbf{x}, m)] + r.\sigma(f(\mathbf{x}, m)), \tag{5.8}$$

where J is the overall objective function, $E_M[f(\mathbf{x}, m)]$ is the expected value, r is the risk factor, and σ denotes the standard deviation. The value of r can vary from -1 to $+1$. $r = 1$ indicates a completely risk-prone attitude, and $r = -1$ indicates a completely risk-averse attitude, and $r = 0$ shows a neutral attitude toward risk.

Semi-variance
This risk measure was first introduced by Makowitz [3]. Semi-variance focuses on the risk associated with the downside of the distribution of an objective function. Based on the trivial approach of deviation criteria, semi-variance is not an averse or coherent measure, but it can provide asymmetric treatment of the objective function distribution [4].

$$Var_+(f(\mathbf{x}, m)) = E[\max(f(\mathbf{x}, m) - E_M[f(\mathbf{x}, m)], 0)]^2, \tag{5.9}$$

$$Var_-(f(\mathbf{x}, m)) = E[\max(E_M[f(\mathbf{x}, m)] - f(\mathbf{x}, m), 0)]^2, \tag{5.10}$$

where $Var_+(f(\mathbf{x}, m))$ represents the variance of the values greater than the mean, and $Var_-(f(\mathbf{x}, m))$ represents the variance of the values smaller than the mean. Therefore, since the focus is on the lower tail of the distribution, Var_- is minimized. Var_- can be combined with the expected profit to form the mean-semi-variance risk measure:

$$J_{MSV} = J_{MO} - \lambda J_{Var_-}, \tag{5.11}$$

where J_{MO} is the expected value of the objective function, J_{Var_-} is the lower semi-variance with the weight factor of $\lambda \in \mathbb{R}$. For more information on the mean-semi-variance please refer to [4].

Value at Risk (VaR)

Value at Risk is one of the widely used risk measures that focuses on the risk of loss. The maximum amount of profit that is expected to be lost with a given probability is known as VaR. Assuming that $f(\mathbf{x})$ is a distributed objective function over multiple realizations, $VaR_\alpha(f(\mathbf{x}))$ is defined in terms of the negative α-quantile:

$$VaR_\alpha(f(\mathbf{x})) = -q_f(\alpha), \alpha \in [0, 1], \tag{5.12}$$

where $q_f(\alpha) = \inf\{z|Prob[f \le z]\rangle\alpha\}$. $q_f(\alpha)$ is defined as the value that the probability of f being lower than $q_f(\alpha)$ in not greater than α. As the definition implies, this methods does not consider the tail of the distribution beyond $q_f(\alpha)$. Consequently, there risk of worst-case outcomes is not eliminated [5].

Conditional Value at Risk (CVaR)

This risk measure was proposed to address the drawbacks of VaR [6]. The difference between the CVaR and VaR is that, unlike VaR, CVaR considers the distribution of the profit beyond the VaR. Therefore, CVaR has a continuous nature with respect to α. To calculate the CVaR, one should compute the expected value of profits beyond the VaR:

$$\mathcal{R}(f(\mathbf{x}, m)) = CVaR(f(\mathbf{x}, m)), \tag{5.13}$$

$$CVaR_\alpha(f) = \frac{1}{\alpha} \int_0^\alpha VaR_s ds, \alpha \in [0, 1]. \tag{5.14}$$

Minimax Regret

The aim of the robust optimization is to end up with decisions that are robust for all available realizations [7, 8]. In risk measures, such as worst-case, semi-variance, CVaR, and VaR, the lower tail of the distribution is taken into account. Therefore, we will obtain solutions that are optimized for worst-case realizations. Most probably, an optimum solution that is obtained using the above mentioned risk measures will not be optimum for the best-case or high-profit realizations. In this manner, if the truth turns out to be from the high-profit realizations, we will regret ignoring high-profit realizations to conduct the robust optimization. Minimax regret risk measure deals with minimizing the maximum value of regret if the selected scenario is far from the truth. Mohammadi and Mohmmadi [9] explained how the minimax regret measure differed from the worst-case scenario.

5.2.3 Proper Selection of Representative Realizations

A large number of geological realizations can be constructed using the uncertain parameters. Even after conditioning the realizations to production data, the number of realizations that match the historical production data is large. Therefore, we have to constrain the number of realizations to decrease the computational cost associated with the robust field development optimization. In this regard, one of the challenges is to select a limited number of realizations which can approximately represent the entire span of the uncertainty in the uncertain parameters.

So far, the common challenges in managing the geological uncertainty in robust FDO have been presented. The following subsection provides a brief review of the recent researches which have tried to resolve these challenges.

5.3 Methods of Uncertainty Management in Robust Field Development Optimization

Uncertainty management in the development phase is investigated by different researchers, including well location optimization [10–15], well control optimization [16–18], joint optimization of well control and location [19], and closed-loop reservoir management [20–22]. This approach although provides reliable results, it associates several challenges which were explained above. Different approaches have been used by various researchers with different risk attitudes to address the challenges associated with robust field development optimization. Some of them have focused on the selection of a representative subset of realizations, and the others have focused on proposing proxy models to substitute reservoir simulators to accelerate the function evaluation process.

5.3.1 Robust Optimization Approaches in Green and Brown Fields

The overall approach to robust optimization (RO) on green and brown fields differ from each other. In green fields, we have limited static and dynamic data, therefore, it is common to generate a large number of geological realizations and conduct RO. On the other hand, as new information becomes available in brown fields, it is common to conduct a history matching step on an ensemble of realizations and use the conditioned realizations for RO. Nonetheless, RO entails utilizing an ensemble of realizations to find the optimal development variables regardless of the field's age. In this regard, some researchers have focused on using all constructed realizations. For example, van Essen et al. [17] considered an ensemble of 100 geological realizations of the standard Egg model to conduct robust optimization of the well controlling parameters to maximize the expected profit. The RO results displayed a much smaller variance, indicating robustness to the geological uncertainty.

Ettehadtavakkol et al. [23] considered a probability distribution for each uncertain parameter and used the Monte Carlo simulation to sample realizations from the probability density function of the uncertain parameters. Therefore, each realization had its probability that was used to calculate the expected profit given in Eq. 5.2.

If the uncertainty range is wide, the number of geological realizations will be large making the robust optimization process computationally demanding. One of the solutions is to reduce the number of realizations by selecting a subset of realizations that can represent the uncertainty in the fullset of realizations.

5.3.2   Different Approaches to Selecting Representative Realizations

Proper selection of a subset of representative realizations can reduce the computational burden associated with RO while keeping most of the range of uncertainty in the fullset. Different selection approaches can be used such as experimental design and selection using static and dynamic properties that are explained below.

5.3.2.1   Experimental Design

Experimental Design (design of experiment) method distributes the simulation runs within the uncertain range of parameters efficiently, thereby, minimizes the number of required runs for studying an uncertain system [24]. In this approach, an experimental design is used to select a number of realizations which provides the maximum information on the entire ensemble. The range of uncertainty for each uncertain parameter is determined and used for sampling a limited number of realizations in the process of the experimental design. In this approach, uncertainty in both the geological variables (e.g., petrophysical properties, net/gross ratio) and engineering variables (e.g., well count) can be changed at the same time. Badru and Kabir [25] used the experimental design to select a representative subset of realizations to capture the geological uncertainty in RO.

Venkataraman [26] suggested a five-step methodology based on the experimental design to capture the range of uncertainty in the recovery factor. The five steps included:

1. Determine the range of each variable affecting the oil recovery.
2. Conduct an experimental design to determine the values of the uncertain parameters at which the recovery is predicted.
3. Predict the recovery of the experiments suggested by the experimental design using a reservoir simulator.
4. Develop a generalized multivariate correlation to predict the recovery.

Table 5.1 Uncertain variables and their range used for generating the geologic realizations [26]

Variable				Scaled variable		
Fault transmissibility	0.01	0.10	1.00	−1	0	1
Poro-perm constant	0.02	0.0475	0.075	−1	0	1
Permeability barriers	0.01	0.10	1.00	−1	0	1
Gas water contact	10,690	10,700	10,710	−1	0	1
Kv/Kh	0.10	0.32	1.00	−1	0	1

5. Validate the developed correlation.
6. Capture the uncertainty in the recovery by predicting its distribution.

The geological realizations were constructed using the variables in Table 5.1.

After developing the predictive model (step 4), it is validated by comparing the results of the model with the results of simulation (see Fig. 5 in [26]). Then, Monte-Carlo simulation was used to generate the distribution of the quantity of interest. Figure 6 in [26] shows the overall workflow to identify the probability distribution of the quantity of interest using the Monte-Carlo simulation.

Experimental design was also used by Moeinikia and Alizadeh [27]. The uncertain parameters and their ranges are shown in Table 5.2. An approach used in conjunction with experimental design is the use of sensitivity analysis to determine the most effective uncertain parameters on the output. For example, Moeinikia and Alizadeh used the one-variable-at-a-time method to determine the most effective parameters on the cumulative oil production.

Table 5.3 shows the values of the Cumulative Oil Production (COP) for different uncertain parameters. In this table, $+1$, -1, and 0 represent the minimum, maximum, and base values for each corresponding variable [27].

Table 5.2 Uncertain parameters and their range used in [27]

Uncertain parameter	Minimum value	Base value	Maximum value
MULTPV	0.5	1	2
MULTX	0.5	1	2
MULTY	0.5	1	2
MULTZ	0.5	1	2
MULTFLT	0.5	1	2
BHP_{inj} (Psia)	4,000	5000	6000
Q_{inj} (Mscf/day)	1000	5000	10,000
P_{ia} (STB/day/Psi)	10	200	400
V_a (STB)	7E + 07	7E + 08	7E + 09
SGCR	0.1	0.2	0.3

Table 5.3 Different uncertain variables and their corresponding COP values according to one-variable-at-a-time design [27]

MULTPV	MULTX	MULTY	MULTZ	MULTFLT	BHP$_{inj}$	Q$_{inj}$	P$_{ia}$	V$_a$	SGCR	COP after 4 years (STB)	COP after 8 years (STB)	COP after 12 years (STB)
+1	0	0	0	0	0	0	0	0	0	15,898,508	33,090,854	47,734,612
−1	0	0	0	0	0	0	0	0	0	12,799,587	24,386,626	33,867,964
0	+1	0	0	0	0	0	0	0	0	13,131,435	29,233,024	41,012,624
0	−1	0	0	0	0	0	0	0	0	13,992,454	28,839,466	39,969,692
0	0	+1	0	0	0	0	0	0	0	14,148,386	28,264,708	41,256,968
0	0	−1	0	0	0	0	0	0	0	13,993,905	27,823,774	39,882,720
0	0	0	+1	0	0	0	0	0	0	15,375,486	31,294,306	44,569,036
0	0	0	−1	0	0	0	0	0	0	13,510,327	26,250,094	36,799,040
0	0	0	0	+1	0	0	0	0	0	14,599,191	29,001,026	40,869,236
0	0	0	0	−1	0	0	0	0	0	14,455,951	28,695,374	40,432,100
0	0	0	0	0	+1	0	0	0	0	14,537,121	28,868,602	40,678,432
0	0	0	0	0	−1	0	0	0	0	14,437,271	28,699,240	40,473,240
0	0	0	0	0	0	+1	0	0	0	14,913,045	30,525,334	43,463,006
0	0	0	0	0	0	−1	0	0	0	14,267,080	27,781,618	38,541,204
0	0	0	0	0	0	0	+1	0	0	14,530,310	28,864,704	40,677,280
0	0	0	0	0	0	0	−1	0	0	14,491,061	28,678,376	40,354,836
0	0	0	0	0	0	0	0	+1	0	14,540,457	28,997,110	41,068,692
0	0	0	0	0	0	0	0	−1	0	14,493,547	28,624,562	40,172,736
0	0	0	0	0	0	0	0	0	+1	14,198,728	28,683,066	40,075,848
0	0	0	0	0	0	0	0	0	−1	14,761,498	30,615,244	44,939,368

Now, the most effective parameters, according to Table 5.3, are MULTZ, MULTPV, SGCR, and Q_{inj}. The effect of each parameter on COP can be determined by Eq. 5.15.

$$
\begin{aligned}
&\text{Effect of the uncertain parameter on COP} \\
&= \frac{\text{main effect of the uncertain parameter on COP}}{\sum |\text{main effect of the uncertain parameter on COP}|},
\end{aligned}
\tag{5.15}
$$

and the main effect of the uncertain parameter on COP is calculated as follows:

$$
\begin{aligned}
&\text{Main effect of the uncertain parameter on COP} \\
&= \text{COP}_{\text{(maximum value of the uncertain parameter)}} - \text{COP}_{\text{(minimum value of the uncertain parameter)}}
\end{aligned}
\tag{5.16}
$$

Figures 2 through 4 in [27] illustrate the effect of each parameter on the 4-, 8-, and 12-year COP. Now, using an experimental design method, various values can be selected for the most effective variables that make the realizations. These realizations are used to estimate the distribution of the COP. Figures 5 through 7 in [27] show the probability density function and cumulative distribution function of the COP for 4, 8, and 12 years.

Dejean and Blanc proposed a method to capture the geological uncertainty in a quantity of interest. For example, the uncertainty in COP can be quantified as follows [28]:

$$
GU = COP(\max) - COP(\min)/((COP(\max) + COP(\min))/2), \tag{5.17}
$$

where GU is the general uncertainty. As a result, GUs at the 4th, 8th and 12th year from the beginning of the simulation are 41%, 60% and 71%, respectively.

One of the drawbacks of the experimental design is that it does not use the full probability distribution of the uncertain parameters to construct the predictive model. Moreover, this approach gives equal weights to all samples meaning that it assumes a uniform distribution for all uncertain parameters [29].

5.3.2.2   Selection using Static and Dynamic Properties

Another approach to select a representative subset of realizations is to cluster them based on static, dynamic or the combination of static and dynamic properties. Clustering is an unsupervised machine learning algorithm in which the algorithm analyzes the structure of the input data to put them into similar clusters [30–32]. This method is used to cluster the geological realizations into separate groups. Therefore, instead of using all realizations, only a few realizations from each cluster are selected. For example, a random realization or the realization corresponding to the centroid of the cluster can be selected from each cluster. One of the important steps in clustering

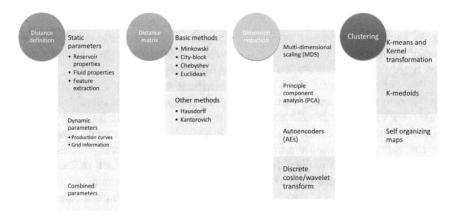

Fig. 5.1   Main steps of the distance-based clustering (DBC) method

is to define a (dis)similarity measure to cluster the realizations. Various distance measures can be used including the Minkowski, City-block, Chebyshev, Euclidean, Hausdorff, and Kantorovich distances. This kind of clustering technique is known as the distance-based clustering (DBC). This distance can be measured between static, dynamic or combined properties of the reservoir. One of the challenges in distance-based clustering is the high dimensionality of the models. For example, a reservoir model may consist of hundreds of thousands to millions of grid blocks complicating the process of calculation of the distance. Therefore, an additional step in DBC is to reduce the dimensionality of the reservoir models. In this manner, first, a dimensionality reduction technique is applied on the realizations, and then a DBC method is used to cluster the realizations. K-means [33], K-medoids [34], and self-organizing maps are some of the DBC methods. These methods are explained in [35]. Figure 5.1 illustrates the main steps of a DBC method.

The distance matrix is constructed by calculating the static, dynamic or combined dissimilarity between the realizations. Static distance can be measured based on static parameters and no simulation is required. However, if the distance is measured between the dynamic response of the realizations, it is classified as the dynamic distance. Among static parameters, permeability is the most used property to calculate the static distance [36–39]. It is also possible to combine several static parameters to calculate the distance [40]. For example, it is possible to combine the permeability, porosity, facies, and water saturation to calculate the static distance. However, it increases the computational and economic expenses. One solution suggested by researchers is to use the average of the properties. For example, instead of using the porosity values of all grids to calculate the static distance, Gross et al. [40] used the mean value of the porosity. However, taking the average value results in a significant loss of information. Therefore, different dimensionality reduction methods, as mentioned above, should be used. These dimensionality reduction methods are discussed in Chap. 4. The distance can be calculated based on the dynamic properties, such as the cumulative oil production, recovery factor, time-of-flight, etc. of two different reservoir models. It is also possible to have a combination of the static and dynamic properties to calculate the dissimilarity distance.

Wang et al. [41] presented a retrospective framework for robust optimization of well locations under geological uncertainties by maximizing the expected net present value (NPV). Their method converted the optimization problem into a sequence of sub-problems with increasing number of realizations. In their method, a number of realizations were selected at each step using the K-means clustering method with an increasing number of realizations from one step to another. Each realization was identified by a vector of static (permeability) and dynamic (cumulative oil production) properties. These vectors were used as the input to the K-means clustering. In the early steps, a fewer number of realizations were selected from the ensemble of realizations. Therefore, the number of optimization iterations was higher. As the algorithm progressed, a higher number of realizations was selected with a fewer number of iterations so that in the last step all realizations were involved with only one optimization iteration. They concluded that the results obtained by the robust optimization were better than the results obtained by only one realization. Their method required to perform clustering at each iteration which can be very computationally demanding if the reservoir model is large.

Using a mixed integer linear optimization (MILP), Rahim and Li [42] proposed a method to select a smaller subset of realizations based on minimizing the Kantorovich distance between the geologic and static properties of the selected subset and the original fullset of realizations. The MILP minimizes the Kantorovich objective function which is defined as follows:

$$\min D_{Kan} = \sum_{i \in I} p_i^{orig} d_i, \tag{5.18}$$

where d_i denotes the cost associated with removing realization i (in other words, the cost of transporting and distributing its probability mass to remaining realizations). One can compute this cost using a weighted sum of the transported probability mass [42]:

$$d_i = \sum_{i' \in I} c_{i,i'} . v_{i,i'}, \quad \forall i \in I, \tag{5.19}$$

where $c_{i,i'}$ is, in fact, the distance between realizations i and i':

$$c_{i,i'} = \sum_k |m_{ik} - m_{i'k}| + \sum_{c,t} \lambda |\theta_{ict} - \theta_{i'ct}|, \forall i, i'. \tag{5.20}$$

In Eq. 5.20, m_{ik} denotes the value of the static property of type k for realization i, θ_{ict} represents the value of the geologic property of type t of grid block c for realization i, and λ is the weight of the geologic properties in the calculation of the dissimilarity function (set to 0.01). Static properties included the average net porosity, the average net permeability, the average net irreducible water saturation, the fraction of the net cell, the net pore volume, and the original oil in-place and net oil in-place. The properties are defined in detail in [42]. The selected realizations were used to calculated the risk-averted expected cumulative oil production as follows:

$$MaxCOP_{risk} = COP_{Expected} - \gamma \sqrt{\sum_{i=1}^{N_R} p_i (COP_i - COP_{Expected})^2}, \quad (5.21)$$

where N_R is the number of realizations, p_i is the probability of realization i, COP_i is the COP of realization i, and γ is the risk-averse factor.

Yousefzadeh et al. [43] used the K-means clustering to automatically cluster 100 equiprobable permeability realizations into 10 clusters. Then, the realizations pertaining to the centroid of the clusters were selected as the representative realizations. Therefore, a subset of 10 realizations was selected based on the static property of the permeability distribution. K-means is one of the most used clustering algorithms whose main steps are described in [44]. The dissimilarity distance between the realizations was the Euclidean distance between the permeability distribution of two realizations which is one the most used and simple distance measures [39, 45–52]. Yousefzadeh et al. constructed a tensor of permeability realizations, A, of size $(nx * nz) \times nz \times 10$ and computed the distance between the realizations and the centroid as follows:

$$d = \| A(:, :, i) - C(:, :, j) \|_F \quad (5.22)$$

where d is the distance between tensors A and C, and $\| ., . \|_F$ is the Frobenius norm between the two tensors. After selecting the representative realizations, they were used to maximize the risk-averse mean-standard-deviation risk measure which was formulated as:

$$J = E_M[f(\mathbf{x}, m)] - \sigma(f(\mathbf{x}, m)). \quad (5.23)$$

This is the same as Eq. 5.8 in which r is taken as -1. After optimization, they captured the geological uncertainty by computing some statistical measures including the mean, standard deviation, cumulative distribution function, p10, p50, and p90 of the objective function. Their results showed that the static clustering method was unable to capture the range of uncertainty in the full set of realizations.

Yousefzadeh et al. [43] also use a dynamic measure to evaluate and group the geological realizations. In the proposed approach, Diffusive Time of Flight (DTOF) is calculated for each realization using the computationally efficient Fast Marching Method (FMM). FMM is a numerical method that can be used to solve problems in the form of the Eikonal equation [53–56], such as pressure propagation in porous media [57, 58]. Therefore, FMM can be used to provide the dynamic response of the reservoir efficiently without the need for any commercial simulator [58–63]. Then, the number of grid blocks having a DTOF less than a predefined threshold is determined which acts as a quality index for each realization. Then, the cumulative distribution function of the quality index is constructed. Now, one can divide the vertical axis of the CDF to n groups to select a realization from each group. In [43], the closest realization to the realization corresponding to the mean of the upper and lower realizations of each group was selected as the representative realization. Yousefzadeh et al. also sorted the realizations based on the quality index and divided them into 10 groups. From each group, a realization was selected that constructed the reduced subset of realizations. They compared the methods based on the mean, standard deviation, CDF, and a statistical distance which was defined as follows:

$$d_m = \sqrt{(P_{10}^m - P_{10}^{All})^2 + \left(P_{50}^m - P_{50}^{All}\right)^2 + \left(P_{90}^m - P_{90}^{All}\right)^2}, \qquad (5.24)$$

where P_*^m and P_*^{All} are the P-* percentile of the objective function for the reduced subset and all realizations, respectively. According to [43], the dynamic measure to select the representative realizations outperformed the static measure. Also, the statistical approach performed better on the cases where the CDF of the objective function was less curvy, and the sorting approach performed better on the cases that had a curvier CDF. This is because, as the CDF becomes curvier, the statistical approach cannot sample realizations from the full span of the uncertainty since most of the realizations concentrate within a small range [43]. Although their method provided satisfactory results, it was limited to only production wells.

A combination of static (log-permeability) and dynamic (flow-response) reservoir properties was used by Shirangi and Durlofsky [64] to select a representative subset of realizations. The dissimilarity measure was the Euclidean distance between the features of the representative subset and the superset of realizations. The features, in fact, were the static and dynamic parameters. Log-permeability was included in a reduced format. The Principal Component Analysis [65] was used to reduce the dimensionality of the permeability field. As a result, a matrix of weighted static and dynamic features was created as follows:

$$\hat{Z} = \begin{bmatrix} \alpha \hat{Z}_p \\ (1 - \alpha)\hat{Z}_f \end{bmatrix}, \qquad (5.25)$$

where $0 \leq \alpha \leq 1$ is the weight factor, in which $\alpha = 0$ corresponds to a feature matrix composed of only flow-based parameters, $\alpha = 1$ results in a fully permeability-based feature matrix, and $\alpha = 0.5$ gives equal weighting to flow-based and permeability-based features. $\hat{Z}_p$ and $\hat{Z}_f$ are the normalized permeability-based and flow-based quantities. After $\hat{Z}$ was created, K-means clustering was applied to divide the realizations into a number of clusters. Afterwards, to select a realization from each cluster, K-medoids algorithm was applied within each cluster to find the centroid that was the representative realization. Shirangi and Durlofsky repeated the realization selection each time the decision variables (well location and control parameters) changed. They tested three values for $\alpha = 0, 0.5, 1$ and recommended to use $\alpha = 0$

to optimize well controlling parameters, and $\alpha = 0.5$ for pattern-search-based optimization of well configurations. For well placement optimization, no specific value for α was recommended. If the computational cost is of concern, it is suggested to use permeability-based features. However, if it is practical to conduct reservoir simulations, they suggested to investigate the proper value of α.

Static and dynamic properties of the reservoir can be combined to construct a single measure that can be used to calculate the dissimilarity measure between different realizations. Vaseghi et al. [66] made use of the Reservoir Opportunity Index (ROI) and Dynamic Measure (DM) to select the representative realizations in green and brown fields, respectively. ROI is a measure of reservoir quality that is composed of static parameters:

$$ROI = \sqrt{0.0314\sqrt{\frac{K}{\phi}}D_x D_y D_z \phi S_o},\tag{5.26}$$

where K is permeability, ϕ is porosity, D_x, D_y, D_z are grid lengths in the x-, y-, and z-direction, respectively, and S_o is the initial oil saturation.

For brown fields, dynamic measure is used. DM is calculated separately for producers and injectors:

$$DM_{producer} = (P \times K \times K_{ro} \times S_o)/V_o,\tag{5.27}$$

$$DM_{injector} = K/(P \times MR \times V_o),\tag{5.28}$$

$$MR = \frac{K_{rw}/\mu_w}{K_{ro}/\mu_o}.\tag{5.29}$$

In Eqs. 5.27 through 5.29, P denotes the pressure, K_{ro} is the oil relative permeability, S_o is the oil saturation, V_o is the oil velocity, MR is the mobility ratio, K_{rw} is the water relative permeability, μ_w is the water viscosity, and μ_o is the oil viscosity. These maps are used with two goals: finding the sweet spots to shrink the search space of the optimization algorithm, and selecting a representative subset of realizations. To find the sweet spots, at first, these maps are converted to density maps in which the value of each cell is replaced with the average of its second-order immediate neighbors [66]. Then, these maps are used to define the Statistical ROI (SROI) and/or Statistical DM (SDM) maps by averaging the maps over multiple realizations. The statistical maps included the mean and standard deviation of the SROI or SDM. Locations having higher mean and lower standard deviation than a predefined threshold are considered sweet spots that correspond to a higher cumulative oil production.

The representative realizations are selected using the K-means clustering. The number of clusters is determined using the Higher-order Singular Value Decomposition (HOSVD) in which several numbers of clusters are examined and the number of clusters resulting in the lowest reconstruction error is selected. For more details on the HOSVD please refer to Appendix A of [66]. The selected realizations are then used for robust optimization of well locations by maximizing the coefficient of variation (see Sect. 5.2.2). The overall workflow of uncertainty management using the SROI and SDM maps is demonstrated in Fig. 1 of [66].

5.3.3 Decreasing the Computational Cost by Constrained Optimization

Since FDO is an iterative process, its computational cost is directly related to the number of iterations. Thereby, decreasing the number of iterations will result in a lower computational burden. To do so, the optimizer can be constrained to a limited search space so that the local/global optimum is found faster. One common problem is well placement optimization in which constrained search can help the optimizer to find the optimal locations with a fewer number of function evaluations. This solution is not limited to robust or nominal field development optimization.

Quality maps (QM) and reservoir opportunity index (ROI) are two of the most common tools to constrain the search space of the optimizer. QM is a two-dimensional map showing the production potential of each location in the reservoir. This map can be constructed using static, dynamic or the combination of static and dynamic properties of the reservoir. One of the most used quality maps is the one proposed by Ding et al. [67] defined in Eq. 5.30:

$$J_{i,j,k}(t) = \left[S_{o,i,j,k}(t) - S_{or}\right].\left[P_{o,i,j,k}(t) - P_{\min}\right].$$
$$\ln K_{i,j,k}.\ln r_{i,j,k}.h_{woc,i,j,k}.\phi_{i,j,k}.h_{goc,i,j,k} \tag{5.30}$$

where $J_{i,j,k}(t)$ is the productivity potential (PP) at gridblock (i, j, k) at time t; $S_{o,i,j,k}(t)$ is the oil saturation at gridblock (i, j, k) at time t; S_{or} is the residual oil saturation at gridblock (i, j, k); $P_{o,i,j,k}(t)$ is the oil pressure at gridblock (i, j, k) at time t; $P_{\min}$ is the minimum well bottom-hole pressure; $K_{i,j,k}$ is the permeability at gridblock (i, j, k); $r_{i,j,k}$ is the distance from gridblock (i, j, k) to the closest reservoir boundary; $h_{woc,i,j,k}$ is the distance from gridblock (i, j, k) to the closest oil–water contact; $\phi_{i,j,k}$ is the porosity at gridblock (i, j, k); and $h_{goc,i,j,k}$ is the distance from gridblock (i, j, k) to the closest gas-oil contact. The PP is calculated for all gridblocks and the regions with high PP values are considered as sweet spots to drill infill wells (see Fig. 4 in [67]). Therefore, the optimizer does not need to search other regions.

Molina and Rincon [68] proposed an ROI which was a combination of the flow capacity of the formation, mobile hydrocarbon pore volume, and the pressure gradient. The parameters mentioned above were obtained from a numerical reservoir simulator. Varela-Pineda [69] also proposed an ROI which was based on the one proposed by Molina and Rincon. However, both of them required flow simulations to get the required dynamic properties. In a recent work, Yousefzadeh et al. modified the ROI proposed by Molina and Rincon as [1]:

$$ROI = \sqrt{0.0314\sqrt{\frac{K}{\phi}} \times \Delta x \times \Delta y \times \Delta z \times \phi \times (S_O - S_{wc}) \times DTOF \times NTG}.$$

$$(5.31)$$

The modifications are the addition of NTG and $DTOF$ and changing the S_{or} to S_{wc}. In Eq. 5.31, K is the permeability field, ϕ is the porosity, Δx, Δy, and Δz are the grid dimensions in the x-, y-, and z-direction, respectively. S_O is the oil saturation, S_{wc} is the connate water saturation, NTG denotes the net-to-gross ratio, and DTOF is the diffusive time of flight calculated by the FMM which is a dynamic property of the reservoir. Therefore, Yousefzadeh et al. were able to combine the static and dynamic properties of the reservoir without using any flow simulation. They also suggested to used ROI density (DROI) maps in which the effect of adjacent gridblocks is considered in calculating the PP of a specific location. Further, they converted the DROI maps of multiple geological realizations to statistical maps by averaging the DROI maps over the realizations. Then, those maps where used to identify the sweet spots to constrain the search space of the optimizer. Their results showed that the constrained optimization could find the optimal results with fewer function evaluations in comparison with unconstrained optimization. Other researchers have also used the ROI maps to facilitate the field development optimization process that can be found in [66, 70–73]

Constrained optimization is not limited to QMs and ROIs. An example can be found in [62], where the FMM is used to find the drainage volume of the existing production wells. Then, a narrow region around the drainage boundary of the producers is determined. These regions are used as sweet spots to drill injection wells to take the maximum advantage of the injectors. Results showed that the constrained optimization was able to find the optimal location of the injectors with fewer function evaluations.

References

1. Yousefzadeh R, Ahmadi M, Kazemi A (2022) Toward investigating the application of reservoir opportunity index in facilitating well placement optimization under geological uncertainty. J Petrol Sci Eng 215:110709. https://doi.org/10.1016/J.PETROL.2022.110709
2. Capolei A, Foss B, Jørgensen JB (2015) Profit and risk measures in oil production optimization. Undefined. 28(6):214–220. https://doi.org/10.1016/J.IFACOL.2015.08.034
3. Markowitz H (1952) Portofilio selection. J Financ 7(1):77–91. https://doi.org/10.1111/J.1540-6261.1952.TB01525.X
4. Siraj MM, Van den Hof PMJ, Jansen JD (2016) Robust optimization of water-flooding in oil reservoirs using risk management tools. IFAC-PapersOnLine. 49(7):133–138. https://doi.org/10.1016/j.ifacol.2016.07.229

5. Mohammadi M, Ahmadi M, Kazemi A (2020) Comparative study of different risk measures for robust optimization of oil production under the market uncertainty: a regret-based insight. Comput Geosci 24(3):1409–1427. https://doi.org/10.1007/S10596-020-09960-7

6. Rockafellar RT, Uryasev S (2002) Conditional value-at-risk for general loss distributions. J Bank Finance 26(7):1443–1471. https://doi.org/10.1016/S0378-4266(02)00271-6

7. Kouvelis P, Yu G (1997) Robust discrete optimization and its applications. Boston, MA, Springer US. (Nonconvex Optimization and Its Applications). https://doi.org/10.1007/978-1-4757-2620-6

8. Ben-Tal A, Ghaoui L El, Nemirovski A (2009) Robust optimization. Princeton University Press. https://doi.org/10.1201/9781315200323-10

9. Mohammadi SE, Mohammadi E (2018) Robust portfolio optimization based on minimax regret approach in Tehran stock exchange market. J Indus Syst Eng. Iranian Institute of Industrial Engineering; 11:51–62

10. Onwunalu JE, Litvak ML, Durlofsky LJ, Aziz K (2008) Application of statistical proxies to speed up field development optimization procedures. In: Abu Dhabi international petroleum exhibition and conference. Abu Dhabi. https://doi.org/10.2118/117323-MS

11. Hutahaean J, Demyanov V, Christie M (2019) Reservoir development optimization under uncertainty for infill well placement in brownfield redevelopment. J Petrol Sci Eng 175:444–464. https://doi.org/10.1016/j.petrol.2018.12.043

12. Busby D, Pivot F, Tadjer A (2017) Use of data analytics to improve well placement optimization under uncertainty. In: Abu Dhabi international petroleum exhibition and conference. Abu Dhabi. https://doi.org/10.2118/188265-MS

13. Mustapha HM, Dias DD (2018) Well placement optimization under uncertainty using opportunity indexes analysis and probability maps. In: ECMOR XVI—16th European conference on the mathematics of oil recovery. Barcelona. https://doi.org/10.3997/2214-4609.201802212

14. Bagherinezhad A, Boozarjomehry Bozorgmehry R, Pishvaie MR (2017) Multi-criterion based well placement and control in the water-flooding of naturally fractured reservoir. J Petrol Sci Eng 149(May):675–685. https://doi.org/10.1016/j.petrol.2016.11.013

15. Sharifipour M, Nakhaee A, Yousefzadeh R, Gohari M (2021) Well placement optimization using shuffled frog leaping algorithm. Comput Geosci 25(6):1939–1956. https://doi.org/10.1007/S10596-021-10094-7

16. Wang C, Li G, Reynolds AC (2009) Production optimization in closed-loop reservoir management. SPE J 14(03):506–523. https://doi.org/10.2118/109805-PA

17. van Essen G, Zandvliet M, Van den Hof P, Bosgra O, Jansen J-D (2009) Robust waterflooding optimization of multiple geological scenarios. SPE J 14(01):202–210. https://doi.org/10.2118/102913-pa

18. Perrone A, Rossa E Della, Spa E (2015) Optimizing reservoir life-cycle production under uncertainty: a robust ensemble-based methodology. In: SPE reservoir characterisation and simulation conference and exhibition. Abu Dhabi. https://doi.org/10.2118/175570-MS

19. Li A, Feng E, Gong Z (2009) An optimal control model and algorithm for the deviated well's trajectory planning. Appl Math Model 33(7):3068–3075. https://doi.org/10.1016/j.apm.2008.10.012

20. Hu YF, Guo TM (2001) Effect of temperature and molecular weight of n-alkane precipitants on asphaltene precipitation. Fluid Phase Equilib 192(1–2):13–25. https://doi.org/10.1016/S0378-3812(01)00619-7

21. Chittineni S, Godavarthi D, Pradeep ANS, Satapathy SC, Reddy PVGDP (2011) A modified and efficient shuffled frog leaping algorithm (MSFLA) for unsupervised data clustering. In: Communications in computer and information science. vol. 192 CCIS. Springer, Berlin, Heidelberg, pp 543–551. https://doi.org/10.1007/978-3-642-22720-2_57

22. Jansen JD, E SI, Siep P, Douma SD (2009) Closed-loop reservoir management. In: SPE reservoir simulation symposium. Woodlands. https://doi.org/10.2118/119098-MS

23. Ettehadtavakkol A, Jablonowski C, Lake L (2017) Development optimization and uncertainty analysis methods for oil and gas reservoirs. Nat Resour Res 26(2):177–190. https://doi.org/10.1007/s11053-016-9308-1

24. Steppan D, Werner J, Yeater R (1998) Essential regression and experimental design for chemists and engineers
25. Badru O, Kabir CS (2003) Well placement optimization in field development. In: SPE annual technical conference and exhibition. Denver. https://doi.org/10.2118/84191-MS
26. Venkataraman R (2000) Application of the method of experimental design to quantify uncertainty in production profiles. In: Proceedings of the SPE Asia Pacific conference on integrated modelling for asset management. OnePetro, pp 205–212. https://doi.org/10.2118/59422-MS
27. Moeinikia F, Alizadeh N (2012) Experimental design in reservoir simulation: an integrated solution for uncertainty analysis, a case study. J Petrol Explor Prod Technol 2(2):75–83. https://doi.org/10.1007/S13202-012-0023-0/FIGURES/8
28. Dejean JP, Blanc G (1999) Managing uncertainties on production predictions using integrated statistical methods. In: SPE annual technical conference and exhibition. OnePetro. https://doi.org/10.2118/56696-MS
29. Sarma P, Xie J (2011) Efficient and robust uncertainty quantification in reservoir simulation with polynomial chaos expansions and non-intrusive spectral projection. Soc Petrol Eng—SPE Reservoir Simul Symp 2:1170–1180. https://doi.org/10.2118/141963-MS
30. Rokach L, Maimon O (2005) Clustering methods. In: Data mining and knowledge discovery handbook. Springer, Boston, MA, pp 321–352. https://doi.org/10.1007/0-387-25465-X_15
31. Goodfellow I, Bengio Y, Courville A (2016) Deep learning. The MIT Press
32. Nerurkar P, Shirke A, Chandane M, Bhirud S (2018) Empirical analysis of data clustering algorithms. Proc Comput Sci 125:770–779. https://doi.org/10.1016/J.PROCS.2017.12.099
33. Hartigan JA, Wong MA (1979) Algorithm AS 136: A K-means clustering algorithm. Appl Stat 28(1):100. https://doi.org/10.2307/2346830
34. Gordon AD, Vichi M (1998) Partitions of partitions. J Classif 15(2):265–285. https://doi.org/10.1007/S003579900034
35. Kang B, Kim S, Jung H, Choe J, Lee K (2019) Efficient assessment of reservoir uncertainty using distance-based clustering: a review. Energies 12(10). https://doi.org/10.3390/EN1210 1859
36. Lee J, Choe J (2016) Reliable reservoir characterization and history matching using a pattern recognition based distance. In: Proceedings of the international conference on offshore mechanics and arctic engineering—OMAE. 8. https://doi.org/10.1115/OMAE2016-54287
37. Kang B, Choe J (2017) Regeneration of initial ensembles with facies analysis for efficient history matching. J Energy Resour Technol, Transactions of the ASME. 139(4). https://doi.org/10.1115/1.4036382/373455
38. Lee K, Jung S, Lee T, Choe J (2017) Use of clustered covariance and selective measurement data in ensemble smoother for three-dimensional reservoir characterization. J Energy Resour Technol Trans ASME 139(2). https://doi.org/10.1115/1.4034443/373357
39. Kang B, Choi J, Lee K, Choe J (2017) Distance-based clustering using streamline simulations for efficient uncertainty assessmen. In: Proceedings of the 18th annual conference of IAMG. Perth, Australia
40. Sahaf Z, Hamdi H, Mota RCR, Sousa MC, Maurer F (2018) A visual analytics framework for exploring uncertainties in reservoir models. In: VISIGRAPP 2018—proceedings of the 13th international joint conference on computer vision, imaging and computer graphics theory and applications, vol 3. SciTePress. pp 74–84. https://doi.org/10.5220/0006608500740084
41. Wang H, Echeverría-Ciaurri D, Durlofsky L, Cominelli A (2012) Optimal well placement under uncertainty using a retrospective optimization framework. SPE J 17(01):112–121. https://doi.org/10.2118/141950-PA
42. Rahim S, Li Z (2015) Well placement optimization with geological uncertainty reduction. IFAC-PapersOnLine. 48(8):57–62. https://doi.org/10.1016/j.ifacol.2015.08.157
43. Yousefzadeh R, Sharifi M, Rafiei Y, Ahmadi M (2021) Scenario reduction of realizations using fast marching method in robust well placement optimization of injectors. Nat Resour Res 30:2753–2775. https://doi.org/10.1007/s11053-021-09833-5
44. Mannor S, Jin X, Han J, Jin X, Han J, Jin X, Han J, Zhang X (2011) K-means clustering. In: Encyclopedia of machine learning. Springer, Boston, MA, pp 563–564. https://doi.org/10.1007/978-0-387-30164-8_425

45. Lee K, Kim S, Choe J, Min B, Lee HS (2019) Iterative static modeling of channelized reservoirs using history-matched facies probability data and rejection of training image. Pet Sci 16(1):127–147. https://doi.org/10.1007/S12182-018-0254-X/FIGURES/17

46. Simon E, Samuelsen A, Bertino L, Mouysset S (2015) Experiences in multiyear combined state–parameter estimation with an ecosystem model of the North Atlantic and Arctic Oceans using the Ensemble Kalman Filter. J Mar Syst 152:1–17. https://doi.org/10.1016/J.JMARSYS.2015.07.004

47. Park J, Jin J, Choe J (2016) Uncertainty quantification using streamline based inversion and distance based clustering. J Energy Resour Technol Trans ASME 138(1). https://doi.org/10.1115/1.4031446/442584

48. Park K, Caers J (2007) History matching in low-dimensional connectivity-vector space. Petrol Geostati 2007 cp-32–00037. https://doi.org/10.3997/2214-4609.201403075/CITE/REF WORKS

49. Scheidt C, Caers J (2009) Uncertainty quantification in reservoir performance using distances and kernel methods-application to a West Africa deepwater turbidite reservoir. SPE J 14(04):680–692. https://doi.org/10.2118/118740-PA

50. Jung H, Jo H, Kim S, Lee K, Choe J (2018) Geological model sampling using PCA-assisted support vector machine for reliable channel reservoir characterization. J Petrol Sci Eng 167:396–405. https://doi.org/10.1016/J.PETROL.2018.04.017

51. Koneshloo M, Aryana SA, Grana D, Pierre JW (2017) A workflow for static reservoir modeling guided by seismic data in a fluvial system. Math Geosci 49(8):995–1020. https://doi.org/10.1007/S11004-017-9696-8

52. Gross H, Honarkhah M, Chen Y (2011) Offshore gas condensate field history-match and predictions: ensuring probabilistic forecasts are built with diversity in mind. In: Society of petroleum engineers—SPE Asia Pacific oil and gas conference and exhibition 2011. Vol 2. OnePetro, pp 1460–1473. https://doi.org/10.2118/147848-MS

53. Sethian JA (2001) Evolution, implementation, and application of level set and fast marching methods for advancing fronts. J Comput Phys 169(2):503–555. https://doi.org/10.1006/jcph.2000.6657

54. Sethian JA (1996) A fast marching level set method for monotonically advancing fronts. Proc Natl Acad Sci 93(4):1591–1595. https://doi.org/10.1073/pnas.93.4.1591

55. Sethian JA, Vladimirsky A (2000) Fast methods for the Eikonal and related Hamilton—Jacobi equations on unstructured meshes. In: Proceedings of the national academy of sciences, vol 97. Berkeley, pp 5699–5703. https://doi.org/10.1073/pnas.090060097

56. Sethian JA, Popovici AM (1999) 3-D traveltime computation using the fast marching method. Geophysics 64(2):516–523. https://doi.org/10.1190/1.1444558

57. Vasco DW, Keers H, Karasaki K (2000) Estimation of reservoir properties using transient pressure data: an asymptotic approach. Water Resour Res 36(12):3447–3465. https://doi.org/10.1029/2000WR900179

58. Sharifi M, Kelkar M, Bahar A, Slettebo T (2014) Dynamic ranking of multiple realizations by use of the fast-marching method. SPE J 19(06):1069–1082. https://doi.org/10.2118/169900-PA

59. Pouladi B, Sharifi M (2017) Fast marching method assisted sector modeling: application to simulation of giant reservoir models. J Petrol Sci Eng 149:707–719. https://doi.org/10.1016/j.petrol.2016.11.011

60. Pouladi B, Keshavarz S, Sharifi M, Ahmadi MA (2017) A robust proxy for production well placement optimization problems. Fuel 206:467–481. https://doi.org/10.1016/j.fuel.2017.06.030

61. Datta-Gupta A, Xie J, Gupta N, King MJ, Lee WJ (2011) Radius of investigation and its generalization to unconventional reservoirs. J Petrol Technol 63(07):52–55. https://doi.org/10.2118/0711-0052-JPT

62. Yousefzadeh R, Sharifi M, Rafiei Y (2021) An efficient method for injection well location optimization using fast marching method. J Petrol Sci Eng 204. https://doi.org/10.1016/j.petrol.2021.108620

63. Yousefzadeh R, Sharifi M, Rafiei Y, Ahmadi M (2020) Dynamic selection of realizations for injection well location optimization. In: 82nd EAGE Annual conference and exhibition, vol 2020. Amsterdam, European Association of Geoscientists and Engineers, pp 1–5. https://doi.org/10.3997/2214-4609.202011103

64. Shirangi MG, Durlofsky LJ (2016) A general method to select representative models for decision making and optimization under uncertainty. Comput Geosci 96:109–123. https://doi.org/10.1016/j.cageo.2016.08.002

65. Pearson FRS K LIII (2010) On lines and planes of closest fit to systems of points in space. London Edinburgh Dublin Philos Mag J Sci 2(11):559–572. https://doi.org/10.1080/14786440109462720

66. Vaseghi F, Ahmadi M, Sharifi M, Vanhoucke M (2021) Generalized multi-scale stochastic reservoir opportunity index for enhanced well placement optimization under uncertainty in green and brown fields. Oil Gas Sci Technol 76(41). https://doi.org/10.2516/ogst/2021014

67. Ding S, Jiang H, Li J, Tang G (2014) Optimization of well placement by combination of a modified particle swarm optimization algorithm and quality map method. Comput Geosci 18(5):747–762. https://doi.org/10.1007/s10596-014-9422-2

68. Molina A, Rincon A (2009) Exploitation plan design based on opportunity index analysis in numerical simulation models. In: SPE Latin American and caribbean petroleum engineering conference proceedings, vol 3. Society of Petroleum Engineers (SPE), pp 1295–1302. https://doi.org/10.2118/122915-ms

69. Varela-Pineda A, Hutheli AH, Mutairi SM (2014) Development of mature fields using reservoir opportunity index: a case study from a Saudi field. In: Society of petroleum engineers—SPE reservoir characterisation and simulation conference and exhibition, RCSC 2013: new approaches in characterisation andmodelling of complex reservoirs, vol 2. OnePetro, pp 615–625. https://doi.org/10.2118/172231-ms

70. Bernadi B, Silalahi ES, Reksahutama A, Miraza D (2020) Infill wells placement in high water-cut mature carbonate field with simulation opportunity index method. In: Society of petroleum engineers—SPE/IATMI Asia Pacific oil and gas conference and exhibition 2019, APOG 2019. Society of Petroleum Engineers. https://doi.org/10.2118/196388-ms

71. Dossary M, Al-Turki A, Harbi B, Faleh A. Progressive-recursive self-organizing maps PR-SOM for identifying potential drilling target areas. In: SPE Middle East Oil and Gas Show and Conference, MEOS, Proceedings. Vols. 2017-March. Society of Petroleum Engineers (SPE); 2017. p. 2238–2248. doi:https://doi.org/10.2118/183803-ms

72. Abd Karim MG, Abd Raub MR. Optimizing development strategy and maximizing field economic recovery through simulation opportunity index. Society of Petroleum Engineers - SPE Reservoir Characterisation and Simulation Conference and Exhibition 2011, RCSC 2011. 2011;(1):554–559. doi:https://doi.org/10.2118/148103-ms

73. W. S, Ariaji T. A new simulation opportunity index based software to optimize vertical well placements, public hearing of final project presentation. In: Public Hearing of Final Project Presentation. 2015.

Chapter 6
History Matching and Robust Optimization Using Proxies

Abstract As mentioned earlier, one of the challenges in history matching and field development optimization under geological uncertainty is the high computational cost of the process. The majority of the computational burden is associated with numerous reservoir simulations required to calculate the misfit/objective function over a large number of realizations. In addition to reducing the number of geological scenarios, another way to reduce the computational cost associated with history matching (HM) and robust optimization (RO) under geological uncertainty is to use proxy models. Proxy models are simpler models than the full-physics reservoir simulation with ignorable computational costs compared to reservoir simulation. These models can substitute the reservoir simulator to speed up the computation of the misfit/objective function. This chapter is dedicated to introducing proxy models used in history matching and robust field development optimization. Different kinds of proxy models; such as physics-based, non-physics-based, and hybrid proxies; pros and cons of using proxy models; and some example applications of proxy models are introduced.

Keywords Proxy · Physics-based proxy · Data-driven proxy · Non-physics-based proxy · Hybrid proxy · Helper proxy · Machine learning · Response surface

6.1 What is a Proxy Model?

Proxy models, also called surrogate models, meta-models, helper functions, are used to substitute the reservoir simulator to accelerate the evaluation of the objective or misfit function [1–8] during HM or RO. Typically, proxy models are simpler models than the conventional fluid flow models in porous media. Thereby, proxy models solve simpler and computationally less demanding equations to approximate the dynamic output of the reservoir. Proxy models can completely substitute the reservoir simulator during the entire HM or RO process, or they can facilitate the process by preliminary analysis of the production strategies. These models can be built based on the physics of the problem, independent of the physics of the problem, or as a combination of two. Therefore, there are three kinds of proxies explained in the following.

6.2   Different Kinds of Proxies

Proxies can be classified into three main categories:

1. Physics-based proxies,
2. Non-Physics-based proxies,
3. Hybrid proxies.

6.2.1   Physics-based Proxies

Physics-based proxy models are those methods constructed based on the physical laws of the system understudy. Physics-based proxies can be simple analytical formulas or numerical methods. For example, one can use the average production potential of wellblocks to optimize production/injection well placement. On the other hand, numerical methods such as the Fast Marching Method (FMM) can be used to approximate the dynamic behavior of a specific reservoir. FMM is a physics-based proxy model that is able to provide a quick and accurate dynamic response of the reservoir without the need for flow simulation by solving simpler equations to obtain the diffusive time of flight throughout the reservoir.

6.2.2   Non-physics-based Proxies

Non-physics-based proxies, do not consider the governing physics of the systems. Non-physics-based proxies are data-driven methods that aim to discover the hidden relationships between the inputs and outputs of the system. The learning process is done through observing a large number of input–output pairs. Supervised machine learning methods fall into this kind of proxy models. Having a dataset of input–output pairs, we can develop a regression model to predict the output of new inputs. Figure 6.1 show a typical example of a univariate regression.

The regressive model can be a linear regression, support vector regression (SVR) [9], XGBoost [10], multi-layer perceptron (MLP) [11], convolutional neural network (CNN) [12], long short-term memory [13], etc. The inputs to the regressive model can be the decision variables (e.g., well locations, permeability distribution, oil saturation, well-pair distances, etc.) and the output(s) can be the cumulative oil production, net present value (NPV), recovery factor, etc. This kind of proxy models require a prior training of the model before it can be used for prediction. For example, a data-driven proxy can take the well locations, oil saturations at well locations, and well-pair distances to predict the oil production or NPV during a well placement optimization process. Another example is to take the permeability distribution and well locations to predict the production/injection data during history matching.

Fig. 6.1 An example of a univariate regression

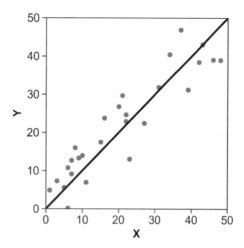

6.2.3   Hybrid Proxies

Hybrid proxy models take benefit from both governing physics of the system and data-driven methods. In other words, a data-driven proxy can take the output of a physics-based proxy as input to estimate the ultimate output. For example, FMM can be used to calculate the diffusive time of flight every time the well locations or the permeability map changes to use it as input to a neural network to predict the production/injection data within robust field development optimization or history matching, respectively. Hybrid proxies have the advantage of incorporating the governing physics of the system to data-driven models to achieve more reliable and case-specific results.

6.3   Application of Proxies in History Matching and Robust Optimization

In this subsection, some of the applications of different proxy models that can completely substitute the reservoir model in history matching and robust field development optimization are presented. Although only a limited number of examples is provided, the implementation process of other proxies is the same as the ones presented here.

6.3.1   Proxies used to Substitute Reservoir Simulators

In this approach, the reservoir simulator is completely substituted with a simpler proxy model to evaluate the objective/misfit function during RO or HM. For example,

Pouladi et al. [7] used the FMM to approximate the geometric bottom-hole pressure drop of the wells in the context of well placement optimization. They formulated the objective function as the bottom-hole pressure drop and related it to the NPV. They assumed that well locations with a lower pressure drop correspond to higher NPVs. Therefore, there was no need for reservoir simulator to evaluate the objective function. For more details on the application of the FMM to approximate the bottom-hole pressure drop of the wells please refer to [7]. This method has shown promising results in predicting the bottom-hole pressure of the wells (see Fig. 3 in [7]). Taking advantage of the new objective function, Pouladi et al. used the FMM to optimize the location of vertical wells on three synthetic models and showed promising results in comparison with optimization using only the reservoir simulation (see Figs. 8 and 10 in [7]). This is while the computational cost of optimization by FMM was orders of magnitude lower than that of optimization by simulator (see Fig. 11 in [7]). Their work showed that the FMM is a robust proxy model that can reliably substitute reservoir simulation to obtain the dynamic response of the reservoir.

Another physics-based proxy is to use objective functions with analytically acquired variables to be evaluated. Therefore, there will be no need for reservoir simulation. An example of an analytical proxy can be found in [14]. Chen et al. proposed an analytical formula-based objective function to optimize well locations, which completely substituted the reservoir simulation [14]. The proposed objective function was a representation of the displacement balance degree of the reservoir, which was derived from the physics of fluid flow in porous media and material balance. The proposed analytical formation is given in Eq. 6.1:

$$\min SD_{DVP} = \sqrt{\frac{1}{N-1}\left[\sum_{i=1}^{N}\left(|DVP_i| - \overline{|DVP_i|}\right)^2\right]}, \qquad (6.1)$$

where, DVP_i is the displacement vector parameter, and SD_{DVP} is the standard deviation of DVP between the injector and producer. An optimal well placement leads to balanced and more displacement of oil from the reservoir to wells. DVP is a reflection of displacement quality of a development plan, which represents the water saturation of injection-production region considering the permeability vector. In other words, the analytical formula of DVP can be obtained by incorporating the permeability vector into water saturation expression. The details of DVP and its derivation can be found in Appendix A of [14]. The formation of DVP differs for different relative permeability curves (Table 1 in [14]). If the distribution of DVP is more uniform in the injection-production region, the displacement is more balanced. This means that the SD_{DVP} tends to zero. On the other hand, if DVP is distributed extremely non-uniformly, SD_{DVP} will tend to 1. Therefore, the objective of the well placement is to minimize Eq. 6.1. Chen et al. showed that using the analytical formula-based objective function significantly decreased the time required for the optimization process.

One of the most used non-physics-based proxy models in the course of HM and RO is the Artificial Neural Network (ANN). An ANN takes benefit from training examples which are pairs of inputs and outputs to train the proxy model. After learning the relationship between the inputs and outputs, the trained ANN can be used during HM or RO to predict the output associated with new inputs. In history matching, inputs are the matching parameters, such as permeability, porosity, etc., and the output is the production data. One of the simplest architectures of the ANN is the multi-layer perceptron (MLP) [11] which is widely used both in the history matching and field development optimization phases for production forecasting [5, 15–17]. MLP consists of three types of layers including the input, hidden, and output layers. Figure 6.2 shows the schematic of an MLP with one hidden layer. Each layer is consisted of several processing units called neurons. Each neuron is the result of the linear combination of all neurons in the previous layer passed through a non-linear function, known as the activation function. Activation function is a non-linear function, such as the hyperbolic tangent (tanh), sigmoid, Rectified Linear Unit (ReLU), etc. which is used to capture the non-linearity in the input variables. In Fig. 6.2, w_{xh} denotes the weights mapping inputs (x) to hidden features, and w_{ho} denotes the weights mapping the hidden features to the outputs (O). This proxy can completely substitute the reservoir simulator by predicting the production/injection data or even the value of the objective/misfit function, or it can be used as a guide to the optimizer.

Bruyelle and Guérillot [18] employed an MLP to history matching of the Brugge case. They trained the network with 761 samples suggested by Box-Behnken experimental design as well as 210 random samples. Their method consisted of two steps:

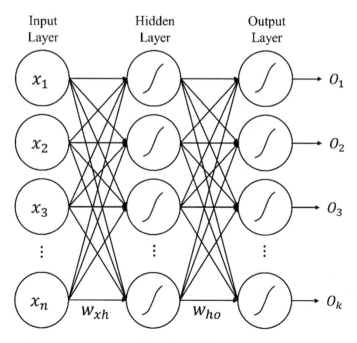

Fig. 6.2 Architecture of a three-layer MLP with one hidden layer

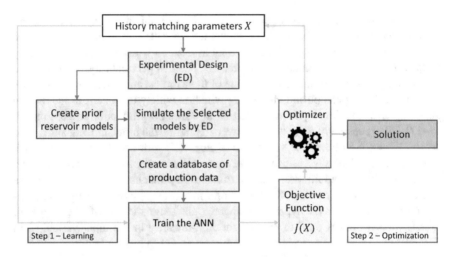

Fig. 6.3 Overall workflow of the history matching methodology using ANN: (1) training the ANN, (2) using the trained ANN to minimize the misfit (objective) function [18]

1- learning, and 2- optimization. In the learning phase, the network is trained using the samples, and in the optimization phase the network is used to minimize the objective function of the history matching process. Figure 6.3 shows the flowchart of the method used by Bruyelle and Guérillot [18]. In the end, the solution is checked with the solution of the reservoir simulator. The authors compared the results of the proposed methodology with the results of conventional proxy models including the response surface and kriging and concluded that the MLP performs better than the conventional history matching proxies.

In addition to MLPs, convolutional neural networks and Long Short-Term Memory (LSTM) are two ANN architectures used for production forecasting during HM or RO. LSTM is generally used for production forecasting by only analyzing historical production data as sequential data. Thereby, it allows us to forecast future performance of the reservoir without the need for any reservoir model and reservoir simulation [13, 19–21].

CNN is a data-driven model that can be used as a non-physics-based proxy model to completely substitute the reservoir simulator. For example, CNN is applied to determining the location of the wells in robust optimization by Kwon et al. [22]. The overall objective function was to maximize the expected cumulative oil production, Q_E^{NN}. The method follows a six-step procedure as follows:

1. In the first step, N_r realizations ($N_r \ll N_r^{\max}$) are used to conduct reservoir simulations to prepare the training, validation, and test datasets; where $N_r^{\max}$ is the total number of permeability realizations. The input of the network is the data regarding the near-wellbore permeability and the output is cumulative oil production. Location of a well in the permeability input data is at its center (see Fig. 2 in [22]).

2. After collecting the training, validation, and testing datasets, the network is trained. Two common measures to measure the performance of regressive models are loss function, J, and correlation coefficient, R. Loss function measures the dissimilarity between the outputs of reservoir simulation and the network. An example of the loss function is the root mean squared error (RSME) with L_2 regularization to prevent overfitting [23]:

$$J = \text{RMSE} + L(\text{W}) = \sqrt{\frac{1}{N_{\text{data}}} \sum_{i=1}^{N_{\text{data}}} \left(Q_o^{RS} - Q_o^{NN}\right)_i^2} + \frac{\alpha}{2}\|\text{W}\|^2, \quad (6.2)$$

where Q_o^{RS} and Q_o^{NN} are the cumulative oil production from reservoir simulation and the neural network, respectively. N_{data} is the number of data points, W is the network's weights, and α is the regularization coefficient which determines the degree of regularization.

Coefficient of correlation is used to measure how much the predictions from the network, Q_o^{NN}, are close to the predictions of the full-physics reservoir simulator, Q_o^{RS}. The formulation of R based on [24] is:

$$R = \frac{N_{\text{data}} \left(\sum_{i=1}^{N_{\text{data}}} Q_o^{RS}i \, Q_o^{NN}i\right) - \left(\sum_{i=1}^{N_{\text{data}}} Q_o^{RS}i\right)\left(\sum_{i=1}^{N_{\text{data}}} Q_o^{NN}i\right)}{\sqrt{N_{\text{data}} \sum_{i=1}^{N_{\text{data}}} Q_{oi}^{RS2} - \left(\sum_{i=1}^{N_{\text{data}}} Q_o^{RS}i\right)^2} \sqrt{N_{\text{data}} \sum_{i=1}^{N_{\text{data}}} Q_o^{NN}i - \left(\sum_{i=1}^{N_{\text{data}}} Q_o^{NN}i\right)^2}}.$$
$$(6.3)$$

3. When a well-trained network is obtained, it can be used to predict the cumulative oil production of new well locations on all realizations (N_r^{max}) for robust optimization.
4. In this step, we can rank the predictions in a descending order. Then, some high-, medium-, and low-rank predictions are selected and compared with the predictions of full-physics reservoir simulation. This step is only for double checking the correlation between the results of the CNN and the reservoir simulation.
5. Based on the results of CNN, two statistical maps, including the mean and standard deviation maps of the cumulative oil production over all realizations are generated. The mean map is used as a productivity potential map. Therefore, by calculating Q_E^{NN} at every feasible well location, locations with a high productivity potential can be determined.
6. Now, a production well is placed at each qualified location from step 4 and the cumulative oil production is predicted using full-physics reservoir simulation for all N_r^{max} realizations to calculate Q_E^{RS}. The optimal well location, $\mathbf{u}^*$ is determined based on the highest Q_E^{RS} if the relative error ε is satisfactory:

$$\varepsilon(\mathbf{u}) = \left|Q_E^{NN}(\mathbf{u}) - Q_E^{RS}(\mathbf{u})\right|/Q_E^{RS}(\mathbf{u}) \times 100(\%). \quad (6.4)$$

In this manner, the optimal well location is identified based on Q_E^{RS}. If a satisfactory value is not obtained for ε, the network is re-trained to get an acceptable performance.

Figure 6.4 shows the overall steps of using the CNN as a proxy model to determine the optimal location of the wells under geological uncertainty (robust optimization). For more information on the reservoir model and the structure of the CNN please refer to [22]. Kwon et al., compared the results of the CNN with those of the multi-layer perceptron and concluded that the CNN performed better. Other applications of CNN to well placement optimization can be found in [25, 26].

Razak and Jafarpour [27] employed a two-step, fully data-driven methodology based on CNNs to history matching of reservoir models. Uncertainty was considered in geological scenarios and in the permeability distribution in each geological scenario. For each geological scenario 500 realizations were generated and simulated to get their production data. Then, each production dataset was labeled based

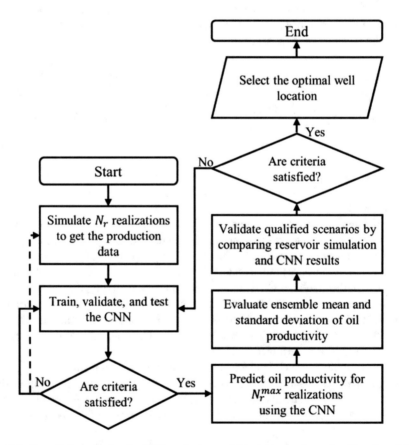

Fig. 6.4 Overall steps of using the CNN as a proxy model to determine the optimal location of the wells under geological uncertainty (robust optimization) [22]

on the geological scenario it was obtained from to build a set of production data and geological scenario labels. Then, a CNN classifier was built to determine the probability of the production data to belong to each geological scenario. This formed the first step of the proposed methodology. Afterward, in the second step, some random realizations were selected from each geological scenario based on the probabilities obtained from the classification phase to train a regression model which could predict the realization associated with the input production data. Note that a dimensionality reduction (PCA) was performed on the production data and model realizations a priori, and then they were used to train the regression model. Figure 6.5 shows the overall workflow of the proposed methodology by Razak and Jafarpour [27]. The advantage of the CNN over conventional history matching methods is that it does not require the probability density function (PDF) of the inputs to be Gaussian [27].

LSTM [28] is a kind of recurrent neural network (RNN) that was introduced to resolve the issues associated with conventional RNNs including the vanishing/exploding gradient and capturing the long-term dependencies in sequential data, such as speech recognition, text processing, etc. LSTM creates self-loops and takes benefit from three gates, including the forget gate, input gate, and output

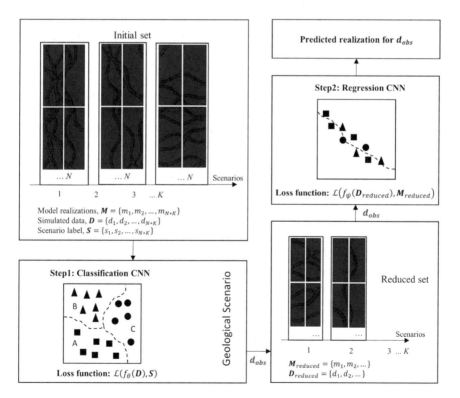

Fig. 6.5 Two-step methodology of history matching using CNNs. Figure produced based on [27]

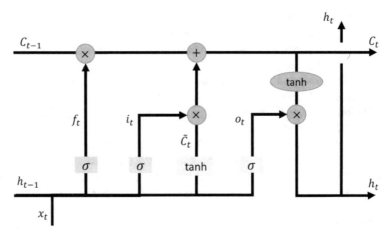

Fig. 6.6 Block diagram of an LSTM cell (processing unit)

gate, which are used for removing and adding information to the cell state. In the self-loops, cells are connected recurrently to each other where information flows between the cells. Cell state is the building block of the LSTM that acts like a conveyor belt which carries information from one unit to the next. Figure 6.6 shows the block diagram of an LSTM processing unit. The forget, input, and output gates as well as the cell state are shown in this figure. Cell state is updated through some addition and multiplication operations, and then is used to calculate the hidden state. If the output of the forget gate is zero, it will remove the information in the cell state; whereas, if the output is one, all information in the cell state is kept. Input gate decides what new information to add to the cell state. If the output of the input gate is zero, no new information is added to the cell state, and if its output is one, all new information is added to the cell state. Using a similar procedure, the output gate controls the output of the cell. For more information about the LSTM please refer to [29].

Azamifard et al. [30] employed the LSTM to predict the dynamic response of the reservoir, including the oil production for 11 time-steps based on permeability realizations. In this approach, permeability realizations were fed into the network in a row-by-row format. Therefore, each row was a sequential vector of permeability data. Permeability realizations were generated using two MPS methods, MS-CCSIM and Graph-cut. However, realizations generated only by the MS-CCSIM method were used to train the network, and the realizations generated by the Graph-cut method were used to evaluate the performance of the designed proxy at predicting the dynamic response of the reservoir based on unseen permeability realizations. Then, the realizations having production data with the lowest difference with actual production data were selected. This ensemble can be used as the initial ensemble of history matching methods such as the ES-MDA, EnKF, etc. Figure 6.7 shows the flowchart of history matching using the LSTM.

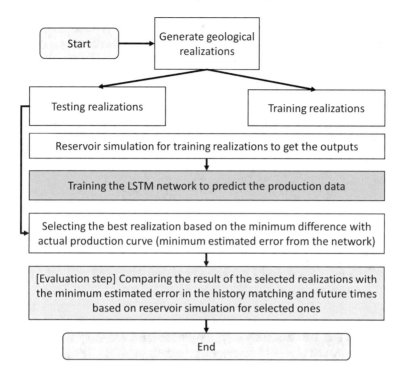

Fig. 6.7 Flowchart of history matching using LSTM

Another application of the LSTM is forecasting the future production of a reservoir based on the historical production data. Therefore, we can construct a non-physics-based proxy to completely substitute the reservoir simulator in forecasting the future behavior of the reservoir, which can be used as a quick tool to analyze the future performance of the reservoir under an ongoing development strategy without the need for any reservoir model. For example, authors in [31] employed an LSTM to forecast the oil production rate for a span of two years by taking benefit from very limited historical production data obtained during the early stages of production. The proposed LSTM technique was able to predict the production in a situation where approximately 70% of the total Estimated Ultimate Recovery (EUR) was recovered in a short period of time. Sagheer and Kotb [13] also used a deep LSTM model to forecast the oil production rate and compared the results with other data-driven and conventional methods, such as the decline curve, auto-regressive integrated moving average (ARIMA), deep gated recurrent unit (DGRU), and vanilla RNN. They concluded that the deep LSTM outperformed the other methods.

There are also conventional methods other than machine learning methods which can be used as proxy models for production forecasting, such as the decline curve analysis, and ARIMA. DCA is a fast and simple empirical method that was initially introduced by Arnold and Anderson [32] and modified by Arps [33]. This technique has been widely used by different authors for production prediction in petroleum reservoirs [34–38]. The main drawback of this method is that there is no way to see the future activities such as well-shut in the prediction task.

One of the most used linear regression models to forecast the stationary time-series data is the ARIMA. It is common to express the model as ARIMA (p, D, q), where p refers to auto regression (AR), q denotes the moving average (MA), and D is the differencing degree. Equation 6.5 provides the mathematical formulation for the ARIMA (p, D, q):

$$\left(1 - \sum_{i=1}^{p} \varphi_i L^i\right)(1 - L)^D x_t = \left(1 + \sum_{i=1}^{q} \theta_i L^i\right)\varepsilon_t, \tag{6.5}$$

where L represents the lag operator, φ_i denotes the parameters of the auto-regressive part, θ_i denotes the parameters of the moving average part, x_t denotes the inputs at time-step t, and ε_t is the error term. These parameters are learned during training the model. Then, the trained model can be used to forecast the oil production without the need for reservoir simulation [39–41].

There are two non-physics-based proxy models other than artificial neural networks. Response surface modelling (RSM) and kriging are two of the non-ANN proxy models that are frequently used for production forecasting in the context of history matching and also field development optimization. In kriging, the outputs of some realizations are simulated and the outputs of other realizations are estimated using a specific kriging method. Response surface relies on simplified polynomial models that model the relationship between the inputs and outputs (response) [42]. Generally, the ultimate goal is to develop a process which relates a response y to some variables $\mathcal{X}_1, \mathcal{X}_2, \mathcal{X}_3, \ldots, \mathcal{X}_n$. Based on [43], the relationship is written as follows:

$$y = f(\mathcal{X}_1, \mathcal{X}_2, \mathcal{X}_3, \ldots, \mathcal{X}_n) + \varepsilon, \tag{6.6}$$

where f denotes the true unknown response function and ε represents the error term. In most second-order models, f involves up to quadratic interactions between the variables. In general, y is given as follows [44]:

$$y = \beta_0 + \sum_{j=1}^{k} \beta_j x_j + \sum_{j=2}^{k} \beta_{jj} x_j^2 + \sum \sum_{i<j} \beta_{ij} x_i x_j + \varepsilon. \tag{6.7}$$

Equation 6.7 can be expressed in a matrix notations [45]:

$$y = X\beta + \varepsilon, \tag{6.8}$$

in Eq. 6.8, X is an $m \times n$ matrix known as the design matrix, y is an $m \times 1$ vector of target values (responses), β is an $n \times 1$ vector of weights, m denotes the number of samples, and n is the number of parameters (features). The error term, ε, is generally assumed to have a normal distribution with a zero mean and a unit variance. As a consequence, we can only model the mean of the response:

$$E[y] = \eta = E[X\beta] = Xb, \tag{6.9}$$

in which b is an unbiased estimate of β. To obtain an estimation of the weights, b, least-squares estimate can be used. In this approach, the weights are estimated by minimizing the discrepancy between the actual and predicted quantities by the RSM, known as the "loss" function. In this regard, the loss function, L, can be formulated as follows:

$$L = \sum_{i=1}^{n} e_i^2 = e^T.e = (y - Xb)^T (y - Xb), \tag{6.10}$$

where e_i is the error between the ith predicted and actual data. Equation 6.10 can be minimized by setting its gradient to zero, $\frac{\partial L}{\partial b} = 0$. Thereby, b is given as follows:

$$b = \left(X^T X\right)^{-1} X^T y. \tag{6.11}$$

Different authors have used RSMs as proxy models to facilitate the history matching process. Monfared et al. [44] combined the response surface with the genetic algorithm to minimize the loss function by matching dynamic parameters, including the oil rate, shut-in pressure, repeated formation test pressure, and water cut.

Nwachukwu et al. [46] suggested to use a combination of physics-based and non-physics-based proxy models to predict the NPV of different production strategies on each realization to compute the expected profit. They suggested to use the Extreme Gradient Boosting (XGBoost) as the non-physics-based proxy model. Like neural networks, XGBoost involves a training step before it can be used for prediction. Their proxy predicted the NPV and other simulator outputs, such as the oil and water production rates, based on the static and dynamic reservoir properties as input. The pairwise connectivities between the wells were also added to the inputs which are a dynamic property forming the physics-based part of the proxy. Pairwise connectivites were represented by the diffusive time of flight which was computed using the FMM. The results on a synthetic reservoir model showed that the proxy was able to predict the responses with a correlation coefficient of 0.99 in comparison with robust optimization using a full-physics reservoir simulator.

6.3.2 Proxies used to Assist Reservoir Simulators

In this subsection, some of the proxy models which can assists the traditional history matching and robust field development optimization methods are presented.

Although only a limited number of examples is provided, the implementation process of other proxies is the same as the ones presented here. It is worth pointing out that the proxy models are common between two approaches (substituting or assisting the reservoir model), while the implementation is different.

For example, the FMM can be used to determine the drainage volume and drainage boundary of production wells to drill the injection wells within the drainage boundary of the existing wells. Yousefzadeh et al. [47] used the same methodology to assist the process of well locations optimization by constraining the search space of the optimizer to nearby regions of the drainage boundary of the production wells. FMM can be used in another way to assist the robust optimization process. This can be done by clustering the geological realizations of a reservoir model based on the dynamic response of the reservoir, which is acquired by the FMM. In this manner, the dynamic response of the reservoir is the diffusive time of flight (DTOF). Yousefzadeh et al. [48] used the same approach to select a representative subset of realizations to be used in the context of RO of well locations. The RO was continued by a reservoir simulator. Using the FMM allows us to cluster the realizations based on a dynamic property without conducting computationally demanding reservoir simulations.

Helper proxies are not limited to FMM. Co-kriging was used in [49] to estimate the dynamic output of high-fidelity models based on the dynamic output of the low-fidelity models. Then, the estimated output data of fine models can be used to build meta-models (proxies). Thereby, by using multi-fidelity meta-modeling to build meta-models at the finest level of resolution by complementing the CPU-time demanding simulations with faster runs at a coarser level. The meta-models were used for history matching in [49].

Another example is proposed by Al-Aghbari and Gujarathi [50] where an evolutionary neural network (EvoNN) was used as a proxy to help the well control optimization process. The EvoNN was trained using the predator–prey genetic algorithm. They formulated the problem as a multi-objective optimization problem in which the non-dominated sorting genetic algorithm II (NSGA-II) was used to simultaneously maximize the cumulative oil production and minimize the cumulative water production. The results of the EvoNN were used as guiding inputs to the NSGA-II algorithm after some iterations to re-initialize the population. As a result, the EvoNN-guided NSGA-II method was able to find better results with a faster convergence compared to the pure NSGA-II algorithm without any assist from the EvoNN [50]. Since the computational cost of the EvoNN was much lower than reservoir simulation, it could evaluate more scenarios to find better solutions than the pure NSGA-II. Therefore, integrating the solutions from the EvoNN to the solutions of the NSGA-II could help the optimizer find better solutions to the problem.

6.4 Pros and Cons of using Proxies in History Matching and Robust Optimization

Using proxy models has its pros and cons. Reducing the computational cost of HM or RO is the greatest advantage of using proxy models. This is accomplished by eliminating the reservoir simulator from the HM or RO process or by assisting the optimization process. As mentioned above, a proxy model can shrink the search space of the optimizer to reduce the number of function evaluations and, subsequently, the number of required simulation runs, which will result in a lower computational burden. The lower time of objective function evaluation enables us to evaluate more development strategies or scenarios in field development optimization or history matching, respectively.

The biggest disadvantage of using proxy models is the lost accuracy of modeling [51]. This means that the modeling uncertainty increases since proxy models make further simplifications regarding the fluid flow in porous media. Physics-based proxy models substitute the full-physics flow equations with simpler equations, such as analytical formula-based objective function [14]. Coming to the FMM, it has two limitations: 1- it cannot consider two-phase and three-phase flow originally, and if one wants to extend it to two- or three-phase flows, it loses its speed. 2- In addition, the FMM suffers from reflections and transmissions of pressure wave at regions with high contrast in permeability distribution which results in an increase or loss in the pressure amplitude [52]. As a consequence, the estimated drainage boundary by the FMM may differ from the reality. Non-physics-based proxy models although do not have the limitations of the physics-based proxies, they need to be trained before they can be used in HM or RO. Training can be problematic and time-consuming since it needs a large number of reservoir simulation runs to prepare the training dataset. In addition, proper training of a data-driven model is challenging since they are prone to overfitting. Also, optimizing the hyperparameters of the models is time-consuming and complicated.

References

1. Byrd RH, Lu P, Nocedal J, Zhu C (1995) A limited memory algorithm for bound constrained optimization. SIAM J Sci Comput 16(5):1190–1208. https://doi.org/10.1137/0916069
2. Eide AL, Holden L, Reiso E, Aanonsen SI (1994) Automatic History Matching by use of Response Surfaces and Experimental Design cp-233–00045. https://doi.org/10.3997/2214-4609.201411186
3. Yousefzadeh R, Sharifi M, Rafiei Y, Ahmadi M (2020) Dynamic selection of realizations for injection well location optimization. In: 82nd EAGE annual conference and exhibition, vol 2020. Amsterdam, European Association of Geoscientists and Engineers 1–5. https://doi.org/10.3997/2214-4609.202011103
4. Yousefzadeh R, Sharifi M, Rafiei Y, Shariatipour SM (2019) An new approach for determining optimum location of injection wells using an efficient dynamic based method. In: 81st EAGE

conference and exhibition 2019, vol. 2019. London, EAGE Publishing BV, 1–5. https://doi.org/10.3997/2214-4609.201900745

5. Shams M, El-Banbi A, Sayyouh H (2019) A novel assisted history matching workflow and its application in a full field reservoir simulation model. J Pet Sci Technol 9(3):64–87. https://doi.org/10.22078/jpst.2019.3407.1545

6. Sharifi M, Kelkar M, Bahar A, Slettebo T (2014) Dynamic ranking of multiple realizations by use of the fast-marching method. SPE J 19(06):1069–1082. https://doi.org/10.2118/169900-PA

7. Pouladi B, Keshavarz S, Sharifi M, Ahmadi MA (2017) A robust proxy for production well placement optimization problems. Fuel 206:467–481. https://doi.org/10.1016/j.fuel.2017.06.030

8. Johnson V, Rogers L (1998) Using artificial neural networks and the genetic algorithm to optimize well field design: phase I final report. California. https://doi.org/10.1016/j.powtec.2017.05.044

9. Drucker H, Surges CJC, Kaufman L, Smola A, Vapnik V (1997) Support vector regression machines. Adv Neural Inf Process Syst 1:155–161

10. Chen T, Guestrin C (2016) XGBoost: a scalable tree boosting system. In: Proceedings of the ACM SIGKDD international conference on knowledge discovery and data mining, 785–794. https://doi.org/10.1145/2939672.2939785

11. Murtagh F (1991) Multilayer perceptrons for classification and regression. Neurocomputing 2(5–6):183–197. https://doi.org/10.1016/0925-2312(91)90023-5

12. LeCun Y, Haffner P, Bottou L, Bengio Y (1998) Object recognition with Gradient-based learning

13. Sagheer A, Kotb M (2019) Time series forecasting of petroleum production using deep LSTM recurrent networks. Neurocomputing 323:203–213. https://doi.org/10.1016/j.neucom.2018.09.082

14. Chen H, Feng Q, Zhang X, Wang S, Zhou W, Geng Y (2017) Well placement optimization using an analytical formula-based objective function and cat swarm optimization algorithm. J Petrol Sci Eng 157(July):1067–1083. https://doi.org/10.1016/j.petrol.2017.08.024

15. Stoisits RF, Crawford KD, MacAllister DJ, McCormack MD, Lawal AS, Ogbe DO (1999) Production optimization at the Kuparuk River field utilizing neural networks and genetic algorithms. In: Proceedings of SPE Mid-continent operations symposium. Oklahoma. https://doi.org/10.2523/52177-MS

16. Aanonsen SI, Eide AL, Holden L, Aasen JO (1995) Optimizing reservoir performance under uncertainty with application to well location. In: SPE annual technical conference and exhibition. Dallas, 67–76. https://doi.org/10.2118/30710-MS

17. Yeten B, Durlofsky LJ, Aziz K (2003) Optimization of nonconventional well type, location, and trajectory. SPE J 8(03):200–210. https://doi.org/10.2118/86880-PA

18. Bruyelle J, Guérillot D (2019) Proxy model based on artificial intelligence technique for history matching—application to brugge field. In: Society of petroleum engineers—SPE gas and oil technology showcase and conference 2019, GOTS 2019. https://doi.org/10.2118/198635-MS

19. Temizel C, Canbaz CH, Saracoglu O, Putra D, Baser A, Erfando T, Krishna S, Saputelli L (2020) Production forecasting in shale reservoirs using LSTM method in deep learning. https://doi.org/10.15530/urtec-2020-2878

20. Song X, Liu Y, Xue L, Wang J, Zhang J, Wang J, Jiang L, Cheng Z (2020) Time-series well performance prediction based on Long Short-Term Memory (LSTM) neural network model. J Petrol Sci Eng. https://doi.org/10.1016/j.petrol.2019.106682

21. Liu W, Liu WD, Gu J (2020) Forecasting oil production using ensemble empirical model decomposition based long short-term memory neural network. J Petrol Sci Eng 189:107013. https://doi.org/10.1016/J.PETROL.2020.107013

22. Kwon S, Park G, Jang Y, Cho J, Chu M Gon, Min B (2020) Determination of oil well placement using convolutional neural network coupled with robust optimization under geological uncertainty. Journal of Petroleum Science and Engineering. 2021:201. https://doi.org/10.1016/j.petrol.2020.108118

23. Dietterich T (1995) Overfitting and undercomputing in machine learning. ACM Comput Surv (CSUR). 27(3):326–327. https://doi.org/10.1145/212094.212114

24. Pearson K (1895) VII Note on regression and inheritance in the case of two parents. Proc R Soc Lond 58(347–352):240–242. https://doi.org/10.1098/RSPL.1895.0041

25. Mousavi SM, Jabbari H, Darab M, Nourani M, Sadeghnejad S (2020) Optimal well placement using machine learning methods: multiple reservoir scenarios. Soc Petrol Eng—SPE Norway Subsurf Conf 2020. https://doi.org/10.2118/200752-ms

26. Kim J, Yang H, Choe J (2020) Robust optimization of the locations and types of multiple wells using CNN based proxy models. J Petrol Sci Eng 193:107424. https://doi.org/10.1016/J.PETROL.2020.107424

27. Razak SM, Jafarpour B (2020) Convolutional neural networks (CNN) for feature-based model calibration under uncertain geologic scenarios. Comput Geosci 24(4):1625–1649. https://doi.org/10.1007/s10596-020-09971-4

28. Hochreiter S (1997) Long short-term memory. Neural Comput 1780:1735–1780. https://doi.org/10.1162/neco.1997.9.8.1735

29. Goodfellow I, Bengio Y, Courville A (2016) Deep learning. The MIT Press

30. Azamifard A, Ahmadi M, Rashidi F, Pourfard M, Dabir B (2020) MPS realization selection with an innovative LSTM tool. J Appl Geophys 179:104107. https://doi.org/10.1016/j.jappgeo.2020.104107

31. Zhang R Han, Zhang L Hui, Tang H Ying, Chen S Nan, Zhao Y Long, Wu J Fa, Wang K Ren (2019) A simulator for production prediction of multistage fractured horizontal well in shale gas reservoir considering complex fracture geometry. J Natl Gas Sci Eng 67:14–29. https://doi.org/10.1016/J.JNGSE.2019.04.011

32. Arnold R, Anderson R (1908) Preliminary report on the Coalinga oil district, Fresno and Kings counties, California. https://doi.org/10.3133/B357

33. Arps JJ (1945) Analysis of decline curves. Transactions of the AIME. 160(01):228–247. https://doi.org/10.2118/945228-G

34. Seshadri J, Mattar L (2010) Comparison of power law and modified hyperbolic decline methods. Soc Petrol Eng—Canadian Unconventional Resour Int Petrol Conf 2010(2):984–1000. https://doi.org/10.2118/137320-MS

35. Can B, Kabir S (2012) Probabilistic production forecasting for unconventional reservoirs with stretched exponential production decline model. SPE Reservoir Eval Eng 15(01):41–50. https://doi.org/10.2118/143666-PA

36. Hu Y, Weijermars R, Zuo L, Yu W (2018) Benchmarking EUR estimates for hydraulically fractured wells with and without fracture hits using various DCA methods. J Petrol Sci Eng 162:617–632. https://doi.org/10.1016/J.PETROL.2017.10.079

37. Zuo L, Yu W, Wu K (2016) A fractional decline curve analysis model for shale gas reservoirs. Int J Coal Geol 163:140–148. https://doi.org/10.1016/J.COAL.2016.07.006

38. Ilk D, Rushing JA, Perego AD, Blasingame TA (2008) Exponential vs. hyperbolic decline in tight gas sands: understanding the origin and implications for reserve estimates using arps' decline curves. In: Proceedings—SPE annual technical conference and exhibition 7:4637–4659. https://doi.org/10.2118/116731-MS

39. Fan D, Sun H, Yao J, Zhang K, Yan X, Sun Z (2021) Well production forecasting based on ARIMA-LSTM model considering manual operations. Energy 220. https://doi.org/10.1016/j.energy.2020.119708

40. Nyangarika A, Mikhaylov A, Richter UH (2019) Oil price factors: Forecasting on the base of modified auto-regressive integrated moving average model. Int J Energy Econ Policy 9(1):149–159. https://doi.org/10.32479/ijeep.6812

41. Mishra AK, Singh S, Gupta S, Gupta S, Upadhyay RK (2022) Forecasting future trends in crude oil production in India by using Box-Jenkins ARIMA. AIP Conf Proc 2481(1):50007. https://doi.org/10.1063/5.0103682

42. Eide AL, Holden L, Reiso E, Aanonsen SI (1994) Automatic history matching by use of response surfaces and experimental design. In: ECMOR IV—4th European conference on the mathematics of oil recovery,. European Association of Geoscientists and Engineers. https://doi.org/10.3997/2214-4609.201411186

43. Myers RH, Montgomery DC, Anderson-Cook CM (2016) Response surface methodology: process and product optimization using designed experiments. Wiley
44. Dehghan Monfared A, Helalizadeh A, Parvizi H (2011) Automatic history matching using the integration of response surface modeling with a genetic algorithm. Pet Sci Technol 30(4):360–374. https://doi.org/10.1080/10916466.2010.483441
45. Dejean JP, Blanc G (1999) Managing uncertainties on production predictions using integrated statistical methods. In: SPE annual technical conference and exhibition. OnePetro. https://doi.org/10.2118/56696-MS
46. Nwachukwu A, Jeong H, Sun A, Pyrcz M, Lake LW (2018) Machine learning-based optimization of well locations and WAG parameters under geologic uncertainty. In: SPE improved oil recovery conference. https://doi.org/10.2118/190239-MS
47. Yousefzadeh R, Sharifi M, Rafiei Y (2021) An efficient method for injection well location optimization using fast marching method. J Petrol Sci Eng 204. https://doi.org/10.1016/j.petrol.2021.108620
48. Yousefzadeh R, Sharifi M, Rafiei Y, Ahmadi M (2021) Scenario reduction of realizations using fast marching method in robust well placement optimization of injectors. Nat Resour Res 30:2753–2775. https://doi.org/10.1007/s11053-021-09833-5
49. Thenon A, Gervais V, Le RM (2016) Multi-fidelity meta-modeling for reservoir engineering—application to history matching. Comput Geosci 20(6):1231–1250. https://doi.org/10.1007/s10596-016-9587-y
50. Al-Aghbari M, Gujarathi AM (2022) Hybrid optimization approach using evolutionary neural network and genetic algorithm in a real-world waterflood development. J Petrol Sci Eng 216:110813. https://doi.org/10.1016/j.petrol.2022.110813
51. Zolfagharroshan M, Khamehchi E (2021) Accurate artificial intelligence-based methods in predicting bottom-hole pressure in multiphase flow wells, a comparison approach. Arab J Geosci 14(4):284. https://doi.org/10.1007/s12517-021-06661-y
52. Wang Z (2018) Asymptotic solutions to the diffusivity equation: validation and field applications. Texas A and M University

Printed in the United States
by Baker & Taylor Publisher Services